# RULES IN THE MAKING

# RULES
# IN THE
# MAKING

*A Statistical Analysis of
Regulatory Agency Behavior*

Wesley A. Magat
Alan J. Krupnick
Winston Harrington

RESOURCES FOR THE FUTURE / WASHINGTON, D.C.

Printed in the United States of America

Published by Resources for the Future, Inc.
1616 P Street, N.W., Washington, D.C. 20036
Books from Resources for the Future are distributed worldwide
by The Johns Hopkins University Press

**Library of Congress Cataloging-in-Publication Data**

Magat, Wesley A.
  Rules in the making.

  Bibliography: p.
  Includes index.
  1. Public administration—Decision making.
  2. Adminstrative agencies—United States—
Management—Decision making. 3. Administrative
procedure—Economic aspects—United States.
  4. Water—Pollution—Law and legislation—United
States. 5. United States. Environmental Protection
Agency—Management—Decision making. I. Krupnick,
Alan J. II. Harrington, Winston. III. Title.
JF1525.D4M33    1986     350.007'25     85-43555
ISBN 0-915707-24-1

The paper in this book meets the guidelines for permanence and durability of the Committee on Production Guidelines for Book Longevity of the Council on Library Resources.

# Resources
## FOR THe FUTURe

RESOURCES FOR THE FUTURE (RFF) is an independent nonprofit organization that advances research and public education in the development, conservation, and use of natural resources and in the quality of the environment. Established in 1952 with the cooperation of the Ford Foundation, it is supported by an endowment and by grants from foundations, government agencies, and corporations. Grants are accepted on the condition that RFF is solely responsible for the conduct of its research and the dissemination of its work to the public. The organization does not perform proprietary research.

RFF research is primarily social scientific, especially economic, and is concerned with the relationship of people to the natural environment—the basic resources of land, water, and air; the products and services derived from them; and the effects of production and consumption on environmental quality and human health and well-being. Grouped into three research divisions—Energy and Materials, Quality of the Environment, and Renewable Resources—staff members pursue a wide variety of interests, including food and agricultural policy, forest economics, natural gas policy, multiple use of public lands, mineral economics, air and water pollution, energy and national security, hazardous wastes, and the economics of outer space. Resident staff members conduct most of the organization's work; a few others carry out research elsewhere under grants from RFF.

Resources for the Future takes responsibility for the selection of subjects for study and for the appointment of fellows, as well as for their freedom of inquiry. The views of RFF staff members and the interpretations and conclusions of RFF publications should not be attributed to Resources for the Future, its directors, or its officers. As an organization, RFF does not take positions on laws, policies, or events, nor does it lobby.

This book is the product of RFF's Quality of the Environment Division, Paul R. Portney, director. It was edited by Charlene Semer and designed by Elsa B. Williams and Martha Ann Bari. The charts were drawn by Arts and Words. The index was prepared by Lorraine and Mark Anderson.

# CONTENTS

TABLES

## FIGURES

# Preface

Efforts to reform the process by which federal administrative agencies carry out the laws passed by Congress are grounded in implicit and only loosely constructed hypotheses about the regulatory process. Further, attempts to test alternative hypotheses have been largely unconvincing because the approaches—primarily case study and anecdotal—yield insights that cannot be confidently generalized. Recently, economists trained in the use of statistical analysis have turned their attention toward the problem of explaining government decision making. The statistical approach has not yet been applied to technical rulemaking by social regulatory agencies, however. The decision making process for these rules is particularly complex and therefore difficult to understand through other approaches.

In this study, we apply statistical techniques to an important and representative example of technical rulemaking, namely the industrial effluent standards set by the Environmental Protection Agency (EPA) pursuant to the Federal Water Pollution Control Amendments of 1972. Our purpose is to develop an improved methodology for exploring how social regulatory agencies issue technical rules, a methodology that derives implicit decision rules from a large number of actual agency decisions. By applying this methodology to industrial effluent standard setting, we demonstrate both its strengths and its weaknesses.

The factors determining the outcomes of EPA's effluent standard-setting process are by no means self-evident. For instance, on December 7, 1973, EPA proposed effluent discharge standards for water pollution from the leather tanning industry. These standards required that by 1977 discharges of biological oxygen demand (BOD) not exceed 40 milligrams per liter (mg/l) of waste water. Four months and two days later, EPA promulgated the final BOD standard for the industry of 102 mg/l. Why was the stringency of the standard weakened by 155 percent between its initial proposal and final promulgation? Why did EPA issue a tighter final standard for the meat packing industry, which produces wastes with similar characteristics to leather tanning, of only 24 mg/l BOD? And why did smaller firms receive weaker regulations?

What factors might have played a role in these outcomes? Did EPA's projection that as many as twenty-one tanning plants would close as a result of the originally proposed standards make a difference in its final decision? Was the Tanners Council's active participation in the leather tanning rulemaking case important? Did the relaxation of the standard result in part from the change in EPA project officers during the rule-making process? Or from the projected costs of compliance? Or do none of these factors systematically affect the stringency of regulatory standards because they are primarily determined by idiosyncratic characteristics of each particular industry's rulemaking process? As an example, was the relaxation of the final leather tanning standard primarily due to the fear that the proposed standard would spur imports of foreign shoes—especially Italian shoes, which a high-level administrator allegedly disliked? Questions about the factors considered by EPA to decide on stringency will be addressed through the use of statistical techniques in an attempt to assess the usefulness of this alternative methodology.

This book reports research funded by a grant to Resources for the Furture, Inc. (RFF) from the Alfred P. Sloan Foundation, to which we are grateful for financial support. Henry M. Peskin of RFF directed the research project and provided invaluable assistance to our efforts. We received valuable suggestions from several other colleagues at RFF, particularly Clifford Russell and Walter Spofford, who provided thorough, critical reviews of various versions of this book.

The idea for this study came from Leonard P. Gianessi, our colleague at Resources for the Future. He suggested that we draw on RFF's extensive experience in analyzing the costs of EPA policymaking to develop a data base describing the EPA rulemaking process. We wish to thank him for his initial suggestion and for his extensive assistance with this project, particularly in helping to organize the massive amount of information on the best practicable technology effluent guidelines into a coherent data base. In addition, his research reconciling the EPA documents with other data sources eased our data collection efforts considerably. Our thanks also go to David Yardas for his major role in collecting and organizing the data and performing computer-related duties. Other research assistants who made valuable contributions to the project included Forest Arnold, Edith Brashares, Joy Hecht, Diane Hill, and James Poles.

We are grateful to the members of the Sloan Project Advisory Committee for their insight and their helpful comments. The members of this committee, which was created especially for this project, included John Brown, Federal Trade Commission; Emery Castle, RFF; Edward Clarke, Office of Management and Budget; Gail Coad, formerly EPA; J. Clarence Davies, Conservation Foundation; David Davis, EPA; Jeffrey Denit, EPA; Lowell Dodge, formerly Consumer Product Safety Commission; Jack Fitzgerald, EPA; Paul Halpern, formerly EPA; Allen Kneese, RFF; Robert Leone, formerly Council of Economic Advisors; Wallace Oates, University of Maryland; John Palmisano, EPA; Paul Portney, RFF; and Clifford Russell, RFF. In addition, participants in several workshops and conferences in which preliminary versions of our work were presented offered useful suggestions. They include Robert Crandall, Paul Downing, A. Myrick Freeman III, Bruce Owen, Sam Peltzman, Jack Stockfisch, Barry Weingast, W. Kip Viscusi, and Lawrence White. Three reviewers of the manuscript as well as our vice president, John Ahearne, and Sally Skillings also provided us with an unusually helpful and thorough set of suggestions, which served as the basis for an extensive revision of the manuscript.

Our thanks go to the dozens of present and former EPA employees who graciously opened their memories and files to us. Without them, the creation of our data base and the understanding we gained about the process of setting effluent standards would have been impossible. Special appreciation also goes to James Patterson of the University of Illinois for his encouragement and his important insights into water pollution control technologies and the regulation of water pollution at EPA. We also would like to express our gratitude to Eileen LesCallett, Mae Barnes, D. J. Curran, and Terri Copeland, who were responsible for the bulk of the typing and secretarial work the project required. Thanks also go to Charlene Semer, who provided scholarly and thoughtful editing, and to Martha Bari who produced the book. Both the Center for the Study of Business Regulation and the University Research Council at Duke University supported the final stages of manuscript preparation.

Wesley A. Magat
Alan J. Krupnick
Winston Harrington

January 1986

# I

# Introduction

Although the effects of industry-specific regulation are well understood, economists and other students of public policy have reached no consensus on how regulatory agencies make rules. Specifically, which factors are important in determining regulatory decisions, and how do agencies trade these factors off against each other in setting their rules? Stated another way, why do regulatory agencies regularly issue standards and other decisions that vary significantly from one regulated party to another? Do distinct factors, such as compliance costs and the quality of information supporting regulatory decisions, systematically explain this variation? Why did the U.S. Environmental Protection Agency (EPA), for instance, set discharge standards for a subcategory of the leather industry that require firms to remove 96 percent of the organic materials in their discharges, whereas the analogous discharge standards for a subcategory of the textile industry mandated only 75 percent removal?

The literature on how agencies set rules is meager. In one of the most exhaustive and often quoted overviews of the regulatory literature, Joskow and Noll (1981) conclude that "Extensive attempts at modelling the behavior of regulatory agencies and regulatory processes have not as yet been forthcoming. . . . Very little research is available on the comparative outcomes of different regulatory institutions."

Despite the limited size of this literature, Joskow and Noll identify research on the behavior of regulatory agencies as one of the most promising directions for regulatory study. In their words, "viewing regulatory commissions as organizations and concentrating on the process of regulatory decision making gives important, useful insights into what

1

is actually happening. The attempts to model and understand regulation from this perspective often give researchers a more complete static and dynamic structural model of regulation rather than just a reduced form. For those interested in incremental policy reform within the context of prevailing institutions, as well as exploring possible institutional alternatives, such structural models are extremely useful for positive policy analysis."

## RESEARCH OBJECTIVES

The purpose of this book is to address directly the question of how regulatory agencies make rules by (1) demonstrating the strengths and weaknesses of a methodology that can provide answers to this question, and (2) applying the methodology to explain the implied decision rules EPA uses to set effluent standards for industrial water polluters based on the "best practicable technology" (BPT), as required by the Federal Water Pollution Control Act Amendments of 1972 (FWPCAA).[1] This particular rulemaking process has served as a model for many subsequent rulemaking procedures designed by EPA, as well as for other social regulatory agencies.[2] It is being used in streamlined form for setting the current round of industrial water pollution standards.[3]

Regulatory agencies, especially social ones, collect and receive a tremendous amount of information that they must process through a contorted sequence of steps to reach decisions about specific rules for particular parties such as plants or firms within an industry category. Often hundreds or even thousands of individuals within an agency participate in the development of a rule, and many different groups, views, and interests are represented within the agency. The deliberations, which can last several years, take place in an agency setting constrained by congressional, executive, and judicial oversight.

As a consequence of this process, in addressing seemingly similar issues in many different contexts, often the standards an agency sets

[1] P. L. 92-500, October 19, 1972. See 33 U.S.C. Section 1251 et seq. Chapter 2 describes the process EPA used to promulgate the BPT standards.

[2] Chapter 2 contains a brief discussion of social regulation.

[3] For details of the process for setting these best conventional technology (BCT) and best available technology (BAT) standards, see "Final Report of Effluent Guidelines Streamlining Study," Program Evaluation Division, Office of Management Systems and Evaluation, February 16, 1982. For detailed definitions of BCT and BAT, see "The Clean Water Act Showing Changes Made by the 1977 Amendments and the 1978 Amendments to Sections 104 and 311." Senate Committee on Environment and Public Works, January 1979.

vary dramatically in stringency among its regulated parties, both for different industries and within the same industry. For example, the 85-decibel noise exposure standard considered by the Occupational Safety and Health Administration (OSHA) would have cost lumber mills $395,000 per worker protected; in contrast, the electrical equipment and supply industry's cost would have been only $39,000 per worker protected. Similarly, OSHA's cotton dust standard of 0.5 milligrams per cubic meter would have imposed notably different costs in two different stages of cotton processing—$56,000 per byssinosis case in yarn preparation, and $22,000 per case in the mill slashing and weaving industry classification.[4] An example from EPA illustrates the same point. Chapter 6 provides data showing that EPA's effluent guidelines water pollution control program required a standard with a marginal cost of only $0.10/kg of Biological Oxygen Demand (BOD) removed for large plants which process chickens, whereas small plants that process ducks (also in the poultry industry classification) were issued a standard with a marginal cost of $3.15/kg of BOD removed.[5]

The stringency of standards also varies dramatically over time within a single rulemaking process. Again referring to the EPA effluent guidelines example, the BOD standards for subcategories in the leather industry were weakened 155 percent from the initial contractor-proposed standard to the promulgated one, while the sugar industry (phase II) standards were strengthened 12 percent over the process. Out of 102 subcategories in the fruits and vegetables industry, twenty-five standards were strengthened and seventy-seven weakened.

The descriptive statistics in tables 1-1, 1-2, and 1-3 further characterize the wide variation in the BPT standards. Table 1-1 summarizes the changes EPA made in the BOD and total suspended solids (TSS) standards for each industry. The changes—which range from −73 percent to 603 percent for TSS and from −12 percent to 174 percent for BOD—usually were accompanied by similar changes for other pollutants. In addition, most changes were made in a direction that weakened the standard.

Table 1-2 further decomposes the changes in the BOD standards. It shows the fractions of subcategories within each of fifteen industries with standards that were strengthened, not changed, and weakened. It also examines these fractions for both the contractor-to-proposed period

---

[4] See Viscusi (1983). Note that these examples are only meant to demonstrate the wide variations in the standards set by agencies, even for the same industries. They are not meant to imply that this variation is necessarily undesirable.

[5] Biological Oxygen Demand (BOD) is a standard measure of the organic waste content of water.

TABLE 1-1.  MEAN UNWEIGHTED PERCENTAGE CHANGE IN MASS EMISSION STANDARDS (kg/kkg) BY INDUSTRY AND RANK

| Industry (ranked by percent change TSS) | Mean % change TSS | N[a] | Mean % change BOD | N[a] | Rank | Mean % change all other[b] | N[a] | Rank |
|---|---|---|---|---|---|---|---|---|
| Leather | 602.9 | 18 | 174.4 | 18 | 1 | 62.9 | 90 | 2 |
| Textiles | 193.0 | 32 | 60.6 | 32 | 2 | 39.9 | 137 | 4 |
| Dairies | 102.3 | 34 | 50.4 | 34 | 5 | 0.0 | 40 | 16 |
| Builders paper | 100.0 | 1 | 20.0 | 1 | 8 | 12.9 | 2 | 8 |
| Asbestos II[c] | 96.0 | 2 | — | — | — | 19.2 | 5 | 6 |
| Cement | 83.3 | 2 | — | — | — | 0.0 | 4 | 16 |
| Fruits & vegetables I | 79.6 | 5 | 56.9 | 4 | 3 | 0.0 | 10 | 16 |
| Fruits & vegetables II | 75.7 | 111 | 51.7 | 113 | 4 | 2.5 | 244 | 14 |
| Phosphates II | 66.7 | 3 | — | — | — | 10.4 | 12 | 9 |
| Sugar I | 60.0 | 2 | 23.2 | 2 | 7 | 0.0 | 4 | 16 |
| Pulp & paper I | 49.3 | 9 | 4.9 | 9 | 12 | 14.0 | 18 | 7 |
| Pulp & paper II | 42.2 | 32 | 19.7 | 32 | 9 | − .43 | 72 | 17 |
| Timber I | 24.2 | 3 | 13.2 | 4 | 10 | 58.1 | 7 | 3 |
| Glass II | 21.2 | 9 | — | — | — | 4.3 | 30 | 13 |
| Meats I | 17.1 | 20 | 36.4 | 20 | 6 | 27.3 | 80 | 5 |
| Grain mills I | 10.7 | 4 | 0.0 | 2 | 12 | 0.0 | 8 | 16 |
| Meats II | 10.1 | 9 | 9.5 | 9 | 11 | 10.2 | 28 | 10 |
| Feedlots | 0.0 | 2 | 0.0 | 2 | 13 | 0.0 | 1 | 16 |
| Glass I | 0.0 | 4 | 0.0 | 1 | 13 | 6.9 | 16 | 12 |
| Asbestos I | 0.0 | 6 | 0.0 | 4 | 13 | 2.0 | 14 | 15 |
| Timber II | 0.0 | 2 | 0.0 | 2 | 13 | − 1.3 | 7 | 18 |
| Grain mills II | 0.0 | 2 | 0.0 | 2 | 13 | 0.0 | 4 | 16 |
| Soaps | − 6.3 | 23 | − 6.3 | 23 | 14 | 8.8 | 110 | 11 |
| Ferroalloys I | − 12.7 | 4 | — | — | — | 386.5 | 29 | 1 |
| Phosphates I | − 50.0 | 2 | — | — | — | − 21.4 | 6 | 19 |
| Sugar II | − 52.0 | 5 | − 12.5 | 4 | 15 | — | — | — |
| Ferroalloys II | − 72.7 | 3 | — | — | — | 0.0 | 6 | 16 |
| Total sample[d] | 92.0 | 346 | 44.4 | 323 | | 27.8 | 984 | |

*Note:* Each industry consists of one or more regulated subsectors. The percentage change in a subsector's standard is calculated by subtracting the standard first recommended by the contractor from the promulgated standard and dividing by the contractor standard. Dashes = not applicable.

*Source:* EPA contractor reports and the *Federal Register*.

[a] N = the number of standards issued.

[b] The mean percentage change for all other pollutants is found by computing the percentage change in the standard from the contractor to promulgated levels for each pollutant other than BOD and TSS in the relevant subcategory, summing these percentage changes over all subcategories in an industry, and dividing by the number of subcategories.

[c] Roman numerals denote rulemaking phases. All of an industry's subcategories were regulated in either the first or the second phase for that industry.

[d] Spearman's Rank Order Correlation Coefficients:

| | | |
|---|---|---|
| TSS-BOD | $r_s = .94$ | $t = 11.5$ |
| TSS-All Other | $r_s = .57$ | $t = 3.4$ |
| BOD-All Other | $r_s = .65$ | $t = 3.1$ |

where

$$r_s = 1 - \frac{6\sum_{1}^{N} d_i^2}{N^3 - N} \text{ and } t = r_s \left(\frac{N-2}{1-r_s^2}\right)^{1/2},$$

and $d$ = the difference in the paired rank for two variables over all observations $N$.

4

TABLE 1-2.  CHANGES IN BPT STANDARDS FOR BOD DURING THE RULEMAKING PROCESS

| Industry | Total number of subcategories | Contractor to proposed | | | | Proposed to promulgated | | | | Contractor to promulgated | | | |
|---|---|---|---|---|---|---|---|---|---|---|---|---|---|
| | | Number of subcategories | | | % change for median subcategory | Number of subcategories | | | % change for median subcategory | Number of subcategories | | | % change for median subcategory |
| | | Strength-ened | No change | Weakened | | Strength-ened | No change | Weakened | | Strength-ened | No change | Weakened | |
| Dairies | 20 | 0 | 20 | 0 | 0 | 5 | 1 | 14 | +35 | 5 | 1 | 14 | +35 |
| Grains | 6 | 0 | 6 | 0 | 0 | 0 | 6 | 0 | 0 | 0 | 6 | 0 | 0 |
| Fruits & vegetables | 102 | 26 | 2 | 74 | +38 | 59 | 6 | 37 | −4 | 25 | 0 | 77 | +33 |
| Seafood | 1 | 1 | 0 | 0 | −8 | 0 | 0 | 1 | +15 | 0 | 0 | 1 | +8 |
| Textiles | 23 | 16 | 0 | 7 | −34 | 3 | 0 | 20 | +70 | 4 | 4 | 15 | +62 |
| Feedlots | 1 | 0 | 1 | 0 | 0 | 0 | 1 | 0 | 0 | 0 | 1 | 0 | 0 |
| Organic chemicals | 7 | 7 | 0 | 0 | < −50 | 3 | 0 | 4 | +45 | 6 | 0 | 1 | < −50 |
| Plastics | 30 | 13 | 3 | 14 | 0 | 5 | 14 | 11 | 0 | 13 | 4 | 13 | 0 |
| Detergents | 23 | 1 | 21 | 1 | 0 | 1 | 22 | 0 | 0 | 2 | 20 | 1 | 0 |
| Leather | 12 | 0 | 0 | 12 | +50 | 0 | 0 | 12 | +50 | 0 | 0 | 12 | +50 |
| Rubber | 3 | 0 | 3 | 0 | 0 | 0 | 3 | 0 | 0 | 0 | 3 | 0 | 0 |
| Timber | 3 | 0 | 2 | 1 | — | 0 | 2 | 1 | — | 0 | 2 | 1 | — |
| Pulp & paper | 35 | 19 | 0 | 16 | −10 | 3 | 0 | 32 | +22 | 9 | 0 | 26 | +16 |
| Builders paper | 1 | 0 | 1 | 0 | 0 | 0 | 0 | 1 | +25 | 0 | 0 | 1 | +25 |
| Meat | 25 | 0 | 20 | 5 | 0 | 0 | 5 | 20 | +27 | 0 | 5 | 20 | +36 |
| All industries | 302 | 89 | 82 | 131 | 0 | 79 | 64 | 159 | +2 | 65 | 48 | 189 | +17 |
| Percent of total | | 29 | 27 | 43 | | 26 | 21 | 53 | | 22 | 16 | 63 | |

*Source:* Resources for the Future Federal Regulations Data Base.

and the proposed-to-promulgated period of the rulemaking process. The table demonstrates that weakening of standards occurred more often between the proposed and final stages of the rulemaking process than between the contractor and proposed stages (53 percent of the subcategories versus 43 percent). The degree to which industries were divided into subcategories also changed from the initial contractor stage of the process to the final stage.

Table 1-3 illustrates the large variation in the number of subcategories into which the firms in each industry were divided. Because each subcategory received a separate set of standards, the number of subcategories indicates the extent to which firms in the same industry were regulated differently. In addition, the table shows that 75 percent of the thirty-one industries listed in the table experienced some change in the number of subcategories into which they were divided, with more subcategories created than eliminated (thirty-one new subcategories in the contractor-to-proposed period, and twenty-two new subcategories in the proposed-to-promulgated period).

What factors account for these wide variations in standards across different industry categories? Economic theory suggests that such factors as the cost of compliance should play a primary role in the decisions about standards. Some political science theories might argue that such politically charged effects as plant closings and unemployment should affect the rulemaking process. Organizational behavior theory identifies technical information about the pollution control technologies as an important factor. We will explore the importance of each of these factors and many others, both in absolute terms and relative to each other, in explaining why some industry categories were issued widely different standards than other categories.

METHOD

This study uses the revealed preference approach to answer the question of how regulatory agencies make rules. This data-oriented approach applies statistical techniques to the outcomes of agency decision making to reveal a set of decision rules that characterize the process the agency used to reach its decisions. This approach allows observers of regulatory agency behavior to infer the agency's preferences for, or trade-offs among, the various factors that enter into its rulemaking decisions. Although agencies rarely develop explicit weights to assign to each factor, analysis of a large sample of rules allows identification of the implicit weights that best describe those trade-offs. The combination of implied weights

TABLE 1-3.   SUBCATEGORIZATION BY INDUSTRY

| Industry[a] | Total number of subcategories at contractor stage | Net change in number of subcategories[b] | |
|---|---|---|---|
| | | C-P | P-R |
| Ferroalloys I | 5 | 0 | − 1 |
| Leather | 7 | − 1 | + 6 |
| Seafoods II | 20 | na | + 4 |
| Seafoods I | 12 | + 3 | + 1 |
| Plastics I | 26 | + 2 | − 2 |
| Dairies | 13 | + 1 | + 6 |
| Timber I | 9 | + 2 | 0 |
| Builders paper | 1 | 0 | 0 |
| Textiles | 7 | + 6 | +10 |
| Fruits & vegetables I | 8 | − 3 | 0 |
| Fruits & vegetables II | 108 | + 7 | −10 |
| Cement | 2 | 0 | + 1 |
| Phosphates II | 7 | − 2 | + 2 |
| Pulp & paper I | 5 | + 4 | − 4 |
| Pulp & paper II | 13 | +13 | + 4 |
| Sugar I | 1 | + 1 | + 2 |
| Meats II | 6 | 0 | 0 |
| Glass II | 7 | + 2 | 0 |
| Glass I | 6 | 0 | 0 |
| Soaps | 21 | + 1 | + 1 |
| Grain mills I | 6 | 0 | 0 |
| Asbestos I | 6 | + 1 | 0 |
| Feedlots | 2 | 0 | + 2 |
| Grain mills II | 4 | 0 | 0 |
| Timber II | 10 | − 2 | − 1 |
| Meats I | 20 | 0 | 0 |
| Plastics II | 15 | − 5 | + 1 |
| Ferroalloys II | 5 | 0 | − 1 |
| Phosphates I | 6 | 0 | 0 |
| Sugar II | 6 | na | + 1 |
| Absestos II | 4 | 0 | 0 |
| Total (net) | 311 | +31 | +22 |

*Source:* EPA contractor reports and the *Federal Register*.
[a] Industries are ranked by the average percent change in standards calculated over all pollutant standards issued for the industry (these data not shown).
[b] C = Contractor stage, P = Proposed stage, and R = Promulgated stage.

and the associated factors is an "implicit decision rule," for it captures the central tendencies that influence the agency's rulemaking decisions. The statistical technique of regression analysis, in concert with theories of rulemaking that turn statistical associations into cause and effect relationships, enables those implicit central tendencies, or weights, to be inferred.

The power of the revealed preference approach lies in its ability to identify both the absolute and relative importance of the factors hypothesized to influence agency decisions. For example, EPA was required to "consider" abatement costs in setting BPT standards, but in practice how did the agency interpret this requirement? Did it equate the marginal costs of control across industries, or did it relax its standards for industries expected to experience many plant closures? The revealed preference approach helps answer this question by identifying which types of costs the agency responded to, be they abatement costs, employment losses, plant closures, or some other form. Further, this approach reveals how the agency implicitly traded off the various forms of cost impacts against its emission reduction goals.

To our knowledge no other studies have used the revealed preference approach to study the standard-setting behavior of social regulatory agencies, such as EPA. The approach does have a small, but important, set of early applications in the regulatory area, however. Joskow (1972) pioneered its use in analyzing the implicit decision rules the New York Public Service Commission used in granting rate adjustments to the electric utilities under its jurisdiction. McFadden (1975, 1976) provided significant advances in both the theory and the application of the method to binary decisions made by governmental agencies, in particular, the freeway-siting decisions reached by the California Division of Highways. Thomas (1979) applied the revealed preference approach to discover the preferences of the Consumer Product Safety Commission (CPSC) members that determined their priority listing for products to undergo CPSC regulatory analysis. More recently, Barke and Riker (1982) reported a statistical case study of the factors that influenced the Interstate Commerce Commission (ICC) in its railroad line abandonment decisions.

McFadden (1976, p. 72) emphasizes that as "applied to the behavior of nonprofit organizations, these methods open the possibility of explaining and improving bureaucratic decision making even when full detail on the command and information structure in the organization is unavailable." Although questionnaires and interviews with government officials might reveal decision rules used for a single regulation or a simple regulatory process, the revealed preference approach remains the only systematic way we know of to understand an agency's decision-making process from outside the agency.[6]

Two other methods of analyzing the behavior of regulatory agencies, case study and informed expert opinion, are important precursors and complements to the revealed preference approach. In fact, one useful way to view our approach is that of a statistical case study. Like all case

[6] As discussed in chapter 7, this approach would not be useful for understanding regulatory processes in which only a small number of decisions are made.

studies, it focuses on one or only a few separate decision-making processes, but it analyzes many more decisions and does so in a systematic manner that allows statistical inferences to be made.

Frequently, analysts use case studies of particular regulatory actions to search for clues about general bureaucratic behavior. A detailed description of the case usually composes the bulk of the study. Suggested policy reforms or general theories of rulemaking rarely emerge from the analyses because the analyst is necessarily cautious about generalizing the results. Generalization requires the assumption that the case under study is "typical"; but the determination of what situation is typical cannot be made on the basis of a single case. Case studies are helpful for understanding the dimensions of the problem, but they are less useful for developing credible regulatory reforms.

For similar reasons, the case study approach is also not well designed to test general hypotheses of agency rulemaking. By focusing on specific actions, the analyst cannot investigate the implications of altering particular factors, such as the White House oversight power or the standards of judicial review, while holding others constant. Testing hypotheses about the influence of specific factors on the regulatory process is difficult, if not impossible, without this capability.

For complicated problems that lack explanations from well-accepted general models in the literature, reformers often turn to informed expert opinion. For example, despite the spectacular growth in the development and use of econometric forecasting models of aggregate economic activity, decision makers still rely heavily upon the judgmental forecasts of "experts." Even though these decision makers lack the ability to test the assumptions behind judgmental forecasts, some prefer them because they take into account additional and more qualitative factors than the econometric forecasts. Of more relevance to this inquiry, legislators, congressional staff members, regulators, and practicing administrative attorneys all develop their own implicit models about the operation of the regulatory process and the effects of various reforms.

The problem is complex, however, and an analyst may become expert in only a few relatively narrow parts of it. Like the blind man describing the elephant by touching each of its separate limbs, each expert may offer a very different set of explanations and reform proposals. Without some method of verifying the representativeness of the experiences of these experts, the basic assumptions they make, and the soundness of their inferences, policymakers are confronted with a large set of different, and often contradictory, theories about how agencies make regulatory decisions and how changing the rulemaking process would affect those decisions.

How does the revealed preference approach overcome the weaknesses of case studies and expert opinions? First, it is specifically designed to allow rigorous statistical tests of its inferences. Second, several applications of the approach to different rulemaking processes will permit judgment on whether the decision rules, or trade-offs, identified for one agency and one rulemaking process are characteristic of other processes as well. Because the decision rules and the methods for inferring them are made explicit, the basis for comparison across different revealed preference studies is much clearer than across several traditional case studies or a collection of expert opinions.

## PLAN OF THE BOOK

Chapter 2 reviews how regulatory agencies make rules, focusing on the informal rulemaking process used to set technical rules. It also compares five important examples of technical and informal rulemaking to determine the extent of heterogeneity within this class of rulemaking processes. Chapter 3 describes in more detail one of these five processes, EPA's procedures for setting its industrial effluent guidelines. It provides the institutional background necessary to understand the statistical analysis of this process.

Chapter 4 briefly reviews the literature relevant to understanding regulatory decision making and then develops a formal model of the standard-setting process based on what has been called the external signals theory. Although of some interest in its own right, this model primarily serves as a mechanism for generating hypotheses that we later test with data from the BPT rulemaking process. The chapter also develops hypotheses from theories about the internal flow of information within an organization. Both of these types of hypotheses help organize the data collection and econometric modeling problems discussed in the next chapter, as well as the statistical hypothesis tests and results presented in chapter 6.

Chapters 5 and 6 are the heart of our research. Chapter 5 discusses the methodological issues posed by the application of our approach, emphasizing the generic problems of developing suitable data for the hypothesis tests and of structuring the econometric model used for the data analysis. We raise the methodological problems likely to be encountered and suggest practical ways of resolving them. Chapter 6 presents the analysis and results from our study of EPA rulemaking in the BPT process. We use regression analysis and other quantitative evidence to test the validity of our general hypotheses as explanations of the implicit decision rules EPA used in setting its effluent guidelines. This chapter also compares our statistical results to those suggested by several case studies of effluent standard setting. Chapter 7 concludes with an analysis of the strengths and weaknesses of applying the revealed preference approach to understanding how social regulatory agencies make rules and an agenda of further research questions.

## REFERENCES

Barke, Richard P., and William H. Riker. 1982. "A Political Theory of Regulation with Some Observations on Railway Abandonments," *Public Choice* vol. 39, pp. 73–106.

Joskow, Paul. 1972. "The Determination of the Allowed Rate of Return in a Formal Regulatory Hearing," *Bell Journal of Economics and Management Science* vol. 3, no. 2 (Autumn), pp. 632–644.

———, and Roger G. Noll. 1981. "Regulation in Theory and Practice: An Overview," in Gary Fromm, ed., *Studies in Public Regulation* (Cambridge, Mass., MIT Press).

McFadden, Daniel. 1975. "The Revealed Preferences of a Government Bureaucracy: Theory," *Bell Journal of Economics* vol. 6, no. 2.

———. 1976. "The Revealed Preferences of Government Bureaucracy: Empirical Evidence," *Bell Journal of Economics* vol. 7, no. 1.

Thomas, Lacy G. 1979. "Revealed Bureaucratic Preferences and Priorities of the Consumer Product Safety Commission." Working paper (Urbana, Ill., Department of Economics, University of Illinois, December).

Viscusi, W. Kip. 1983. *Risk by Choice: Regulating Health and Safety in the Workplace* (Cambridge, Mass., Harvard University Press).

# 2
# Agency Rulemaking

Although the revealed preference approach could be applied to most bureaucratic decision making, the present work is most relevant to a narrower class of decisions, namely technical rulemaking by social regulatory agencies. This distinction needs further clarification, both because the words "technical," "rulemaking," and "social" are sometimes defined differently and because the critical aspects of technical rulemaking need to be kept clearly in mind in interpreting the study results.

## SOCIAL REGULATION

Prior to 1970, most federal regulation of any consequence concerned profits of natural monopolies and certain other industries such as natural gas production and distribution, commercial aviation, railroads, and trucking. In contrast, many of the regulations since 1970 have been directed at broader social problems such as health and safety in the workplace, consumer product safety, and environmental quality. To address these social problems, Congress adopted the strategy of creating new agencies and delegating to them broad powers of implementation.

The new social regulation departs from traditional economic regulation in three additional ways: (1) each agency must deal with a wide variety of industries and other interest groups; (2) the regulatory outcomes turn on factual issues that are difficult, if not impossible, to verify; and (3) social regulatory agencies control only one aspect of a firm's operations or a consumer's behavior (for example, the Occupational

Safety and Health Administration [OSHA] is concerned only with workplace hazards), while other agencies, such as the Federal Maritime Commission (FMC), regulate most of the major decisions made by one particular industry.

Consider some of the consequences of these characteristics. The capture theory of regulation proposes that employees of single-industry agencies, over time, begin to identify the interests of the agency with the interests of the industry.[1] This "capture" results from the constant contact of agency employees with industry personnel and also from "revolving door" practices—key employees either have worked for the industry before entering government service or hope eventually to work for the industry.

Unlike single-industry agencies, such regulatory agencies as the Environmental Protection Agency (EPA), OSHA, and the Consumer Product Safety Commission (CPSC) are more difficult for the industries they regulate to capture, simply because they regulate so many of them.[2] Nevertheless, a partial capture is possible. An agency that regulates tens or hundreds of industries and that faces difficult questions of fact must look outside for expertise about any particular industry. Industry holds much of the data and experience necessary to determine the facts and resolve difficult questions. Of course, an agency cannot completely rely on industry to provide, organize, and interpret data because of obvious conflicts of interest. Where gaps in information exist, however, an industry can exercise decisive influence—in effect, capturing the agency through its monopoly of information.

One characteristic of social regulation that has received little comment from students of government regulation—yet has important rulemaking implications—is industry categorization. Often, groups of firms or individuals regulated by the same rule are heterogeneous, and differences among them can cause differences in the impact of a regulation. Under these circumstances, an agency can encounter considerable opposition to a rule even though the rule is acceptable to the majority of those parties it will affect. One solution to this problem is to categorize the parties into several subgroups having one or several common charac-

---

[1] The capture theory is discussed more fully in chapter 4.

[2] Interestingly, before the Reagan Administration, the "revolving door" notion for social regulation worked differently. Many top officials at EPA, for example, were recruited from public interest groups and law firms specializing in regulatory matters. Many others left EPA to join law and consulting firms after their period of government service, contributing to the growing judicialization of the rulemaking process as well as to its dependence on private consultants. Similarly, prior to the early 1980s, OSHA was considered to be heavily influenced by labor rather than any particular industry group.

teristics and then set separate standards for each subgroup. The most familiar form of categorization is probably "grandfathering," in which an agency issues one set of restrictions for existing firms or plants and a different, more stringent set for new entrants.[3] Industry opposition to this practice is understandably low because it acts as a barrier to entry, thereby reducing competitiveness within the industry.

## RULEMAKING

Until recently, federal agencies promulgated regulations according to one of two procedures established by the Administrative Procedures Act (APA) of 1946. The more demanding procedure was formal rulemaking, described in 5 U.S.C. Sections 556 and 557. This procedure typically required a formal, trial-type hearing before an administrative law judge for the purpose of taking evidence related to the matter in question. In this hearing, interested parties had the right to conduct cross-examination. An agency decision could only be based on the record developed in this hearing, and during judicial review, the proponent of the rule had the burden of proving that it was justified by the "substantial evidence of the record."[4]

The second procedure described by the APA—informal or "notice and comment," rulemaking—was designed to be much less rigorous. This procedure, established in Section 553, required only that the agency (1) issue a notice of proposed rulemaking in the *Federal Register*, (2) give interested parties an opportunity to comment on the proposed rule, and (3) include in the promulgated rules "a concise general statement of their basis or purpose." No cross-examination was required, nor did the APA require the agency to construct a record. Even if a record was kept, not until recently was the agency required to base its decision only on the facts of the record.[5]

The scope of judicial review in informal rulemaking was limited to whether an agency action had been "arbitrary and capricious." On review, the courts generally presumed that facts justified the agency action; to overturn a rule, the challenging party had to rebut this presumption. Judicial review of informal rulemaking has been compared to that for

[3] See Leman (1980) for an interesting discussion of grandfathering in different policy contexts.

[4] Private individuals, firms, or, more commonly, the agency itself could propose rules.

[5] The legislative history of the APA suggests that the rules were presumed to have a factual basis, but those facts could be scattered in agency files or in the minds of agency personnel. See Davis (1978, p. 452).

legislating. Courts can only rule on whether legislation is constitutional, not on the quality of the legislation. Similarly, until recently, courts only decided whether an agency's decision could have been reached by considering the facts of the case but not whether the agency reached the best decision, given the facts.

The APA also exempted certain classes of rules from some or all of the procedural requirements of informal rulemaking, including matters relating to public property, loans, grants, benefits or contracts, foreign affairs, interpretative (as opposed to substantive) rules, substantive rules granting exemptions from existing regulations, rules of agency organization or procedure, and situations in which the agency can show good cause that the use of informal rulemaking is contrary to the public interest. The majority of regulations federal agencies issue probably fall into these exempted classes, although few of those regulations have substantive impacts on the groups they regulate.

Today's federal rulemaking procedures are significantly different from the APA model. Two trends are especially noteworthy. First, the use of formal rulemaking for rules of general applicability has declined sharply.[6] This procedure simply has not proved to be very useful for establishing policies applicable to a large number of firms. Second, Congress and the courts have together imposed so many additional constraints on informal rulemaking procedures that informal rulemaking as envisioned by the APA no longer exists. In its place have emerged numerous procedures. Although these procedures are often unique to the particular statute mandating the rules, they are nonetheless similar in most important respects. These procedural changes amount to a "revolution" in notice-and-comment rulemaking, brought about by a series of mutually reinforcing actions of the courts and Congress. Some have called the new procedures "hybrid" rulemaking—neither formal nor informal, but something in between.[7]

[6] Formal rulemaking procedures continue to be used frequently for rules pertaining to particular firms or individuals. Examples include ratemaking procedures by the Interstate Commerce Commission and the Federal Energy Regulatory Commission, licensing decisions by the Federal Communications Commission, and cancellation of pesticide registrations by the EPA under the Federal Insecticide, Fungicide and Rodenticide Act. Even in these situations, formal rulemaking is on the decline. First, the trend toward deregulation is eliminating much federal ratemaking authority. In addition, agencies are beginning to use modified informal rulemaking procedures even for ratemaking. They have been assisted in this shift by recent court decisions and acts of Congress. For example, the Department of Energy Organization Act, 42 U.S.C. Section 7173(c), permits rates established under the Federal Power Act or Natural Gas Act to be conducted through informal rulemaking procedures. See Davis (1978).

[7] "Revolution" is a term originated by Davis (1978). For further discussion of hybrid rulemaking, see Williams (1975).

For their part, the courts often are no longer willing to defer to the expertise and authority of regulatory agencies. The APA required only a "concise general statement of their basis and purpose" as justification for agency actions. Recent court decisions have also required a summary of facts or findings (which may include statements of agency policy) on which the rule is based and, in some instances, have required agencies to respond to adverse comments with factual materials.[8] The courts have also begun to scrutinize the use of facts in support of the rules. The distinction between the "arbitrary and capricious" and "substantial evidence" standards has largely disappeared because the courts have branded failures to consider certain facts in the rulemaking procedures "arbitrary and capricious."[9] Indeed, the courts have on occasion remanded informal proceedings to give subjects an opportunity for cross-examination on specific issues of fact.[10]

These changes in judicial attitudes have had substantial impact on rulemaking procedures. For one thing, agencies must consider comments more seriously than before. In addition, prudence requires that agencies maintain records that document their consideration of issues and comments. Both of these changes have moved informal rulemaking far from its initial analog with the deliberations of a legislative body.

Congress has been as active as the courts in imposing more stringent requirements on informal rulemaking procedures. Since 1968, it has enacted very few statutes that do not include procedural requirements that go beyond Section 553 of the APA. For example, Sections 304(b) and 307(a) of the 1972 Federal Water Pollution Control Act Amendments direct EPA to "consult" with other federal agencies, state and local officials, and other interested parties. Other statutes enacted since 1968 mandate even stricter requirements. The 1976 Toxic Substances Control Act and the 1974 Federal Trade Commission Improvement Act each require that the agency conduct a public hearing, with a limited right of cross-examination. The 1970 Occupational Safety and Health Act also requires hearings, but the statute is silent on cross-examination. Several statutes introduced during the period made explicit the notion of a rulemaking record, and regulations were to be supported by the "substantial evidence in the record."

Although these changes were increasing the procedural safeguards of informal rulemaking, other changes were increasing its applicability. Some agencies began to use informal rulemaking for rules in categories exempted by the APA.[11] The courts also have cut back on some of the

[8] *United States v. Nova Scotia Food Products Corp.* 568 F. 2d 240 (2d Cir. 1977).
[9] *Associated Industries v. Department of Labor*, 487 F. 2d 342 (2nd Cir. 1973).
[10] *International Harvester Co. v. Ruckelshaus*, 478 F. 2d 615 (D.C. Cir. 1973).
[11] See, for example, Department of Agriculture, 36 F. R. 13804.

exemptions, especially the one for interpretive rules and general policy statements. The current state of the law is that such rules are not exempt from informal procedures if the rules have substantial impact on those affected.[12] Most experts in administrative law probably would agree that fewer exemptions are desirable, although they may not agree on where to draw the line.[13] Future changes in the law, whether by Congress or by the courts, will undoubtedly expand the use of informal procedures for exempted categories.

## TECHNICAL RULEMAKING

In recent years, Congress has legislated technical rulemaking by social regulatory agencies. Although still using the APA rulemaking procedures, under technical rulemaking agencies issue rules based upon a specific technology applicable to the regulated parties. "Purely" technical rules are set at a level consistent with the use of a given technology (usually the so-called "best" technology) without explicit regard to whether the regulation advances the goal of the enabling legislation or to the costs the regulation imposes upon the regulated parties. Few, if any, standards are purely technical, because the enabling statutes often require that the standards be set considering costs, energy use, community impacts, and other external factors. In addition, most technical rules mandate standards with levels that are achievable through the use of the best technology and that refer to the agency's background information documents describing those technologies, but they do not require that firms actually install the designated technology. In practice, however, firms often do adopt the recommended technologies because (1) firms lack knowledge of better alternatives (which may not exist); (2) firms can more easily demonstrate their compliance with the standard by using the agency's suggested hardware; and (3) firms can blame the agency for poor advice if they fail to meet the standards.

The use of technical standards is fairly common by authorities at all levels of government. Consider, for example, their familiar use in local building codes. At the federal level, one highly visible proposal for use of such standards is the requirement that new cars be fitted with passive restraints. The OSHA "consensus" standards for workplace safety are another good example. One of these standards requires that hand rails

---

[12] The rationale is that such a procedure is required by "elementary fairness." *Independent Broker-Dealer Trade Association v. SEC*, 442 F. 2d 132 (D.C. Cir. 1971).

[13] Administrative Conference Recommendation 76-7 urges federal agencies to use informal rulemaking for interpretative rules and policy statements when such actions have substantial impact.

be 30 to 34 inches high and at least 2 inches thick if made of hardwood (1½ inches for metal pipe).[14]

Because the Occupational Safety and Health Act of 1970 requires OSHA to set standards that are technologically feasible and that provide workers with an adequate margin of safety, most OSHA standards are not purely technical. Nor are EPA's effluent guidelines for water pollution.[15] The federal Water Pollution Control Act Amendments of 1972 (FWPCAA) directed EPA to set standards reflecting the best practicable technology (BPT) and to set later standards consistent with the best available technology (BAT), although costs were to be considered. The Clean Air Act of 1970 requires new source performance standards (NSPS) to reflect "the degree of emission reduction achievable through the application of the best system of continuous emission reduction which (taking into consideration the cost . . .) . . . has been adequately demonstrated . . . ."[16]

At first blush, it might seem odd that Congress would resort to technical rulemaking. After all, the benefits and costs of a regulation are central to its success, yet they may not receive much attention or be properly balanced against one another in a technical rulemaking procedure. To many authorities both in and out of Congress, however, this type of regulation promises a quick, easy technological fix to social problems, while also limiting agency discretion. For the most part, the assumption is that disinterested experts at an agency, in the absence of political influence, can unambiguously identify an appropriate technology without undue delay. Many decision makers further assume that technical rulemaking can avoid most of the practical and philosophical difficulties seen in more formal cost-benefit analyses of regulations. Unfortunately, however, the technical approach to rulemaking is an exceedingly complex regulatory process.

## FIVE RULEMAKING PROCEDURES

A comparison of the BPT process, the subject of our empirical work, with four other informal and technical rulemaking processes gives a sense of the heterogeneity among different informal and technical rulemaking processes and presents a sharper picture of the BPT process. In addition, examination of these processes indicates the extent to which the other four processes are amenable to analysis by the revealed pref-

[14] 29 CFR 1910.23.
[15] 29 U.S.C. Section 650 et seq.
[16] P.L. 91-604, December 31, 1970. See 42 U. S. C. Section 1857 et seq.

erence approach. The four processes include two at EPA, one at OSHA, and one at the Consumer Product Safety Commission (CPSC). Like EPA's procedures for setting effluent standards, these four procedures were covered by Executive Orders 12044 and 12291.[17]

The four specific rulemaking processes generate the following types of standards:

1. *Performance standards for new, stationary sources of air pollution.* EPA's statutory authority for promulgating these standards is Section 111 of the 1970 Clean Air Act (CAA) (42 U.S.C. Section 1957 et seq.).

2. *Noise emission standards.* EPA promulgates performance standards for new products under Section 5 of the Noise Control Act of 1972 (42 U.S.C. Section 4901 et seq.).

3. *OSHA health standards.* Pursuant to the 1970 Occupational Safety and Health Act (29 U.S.C. Section 650 et seq.), OSHA, within the Department of Labor, promulgates both performance and design regulations for employers in order to protect the health and safety of workers. OSHA's first standards were "consensus" standards, that is, recommendations of independent standard-setting organizations, which OSHA allowed to become regulations without rulemaking procedures. Since the early adoption of the consensus standards, OSHA has focused its rulemaking efforts on new health and safety standards. In this section, only the health standards are considered.

4. *Consumer product safety standards.* CPSC promulgates regulations—product performance and design standards—to protect the public from hazardous consumer products. CPSC was established and given standard-setting authority by the 1972 Consumer Product Safety Act (15 U.S.C. Section 2051 et seq.).[18]

Table 2-1 summarizes the answers to eighteen questions that compare characteristics of rulemaking across the five processes. Further explanation of these answers follows.

---

[17] "Improving Government Regulations," Executive Order 12044, *Federal Register*, March 24, 1978, and Executive Order 12291, *Federal Register*, February 19, 1981.

[18] At the same time CPSC was given the responsibility for promulgating and enforcing standards under four previously enacted statutes: the Hazardous Substances Act (1960), The Flammable Fabrics Act (1953), the Poison Prevention Packaging Act (1970), and the Refrigerator Safety Act (1954). The procedures the agency used to implement these acts are similar to those used for the 1972 act.

TABLE 2-1. CHARACTERISTICS OF FIVE RULEMAKING PROCESSES

| Characteristics | EPA BPT | EPA NSPS (air) | EPA Noise Control | OSHA (health) | CPSC |
|---|---|---|---|---|---|
| 1. Extent of agency discretion allowed | Much | Much | Much | Much | Much |
| 2. Performance standards issued | Yes | Yes | Yes | Yes | Yes |
| 3. Use of project officers expert in predominant areas | Yes; Engineer | Yes; Engineer | Yes; Engineer | Yes, along with a project attorney; industrial hygiene or toxicology | Yes; Engineer |
| 4. Firms required to submit data | Yes | Yes | No | No | No |
| 5. Contractors used | Yes | Yes | Yes | Yes | Yes |
| 6. Economic and technical analysis performed | Yes | Yes | Yes | Yes | Yes |
| For different offices? | Yes | No | No | Recently, yes | Yes |
| 7. Works closely with industry | Yes | Yes | No | No | Yes |
| Industry has most data? | Yes | Yes | Varies | Yes | Yes |

| | | | | | |
|---|---|---|---|---|---|
| 8. Working groups used | Formal | Formal | Formal | Informal | Informal |
| 9. Interagency review used | Yes | Yes | Yes | Now, yes | Now, yes |
| 10. Influential advisory group | No | Yes | No | Occasionally | Yes |
| 11. Public comments from all interested parties | Almost all from industry, seeking to have guidelines weakened | Usually from industry | Usually from industry | Comments come equally from industry and labor | Both consumers and industries |
| 12. Tight deadlines | Yes | Yes | Yes | Not usually | No deadlines |
| 13. Supporting documents distributed before formal proposal | Yes | Yes | Yes | Sometimes | No |
| 14. Hearings | If asked | If asked | If asked | Yes | Yes |
| 15. Court challenges | Often | Seldom | Usually | Usually | Often |
| 16. Number of major rules (10/1/83) | 87 | 42 | 7 | 9 | 35 |
| 17. Data for statistical approach available | Yes, with difficulty | Yes, with difficulty | Yes | Yes | Yes |
| 18. Categorization used | Extensively | Moderately | Occasionally | Occasionally | Occasionally |

1. *Do the agencies have discretion over the stringency of their rules?* Wide discretion appears to characterize each of the five rulemaking procedures. For each, the enabling statutes are loosely worded, allowing the agencies considerable discretion in interpreting their legislative mandates. The language in the Federal Water Pollution Control Act Amendments is quite vague, requiring that the BPT standards be based on the best practicable technology without precisely defining either "best" or "practicable." New source performance standards (NSPS) are to reflect "the degree of emission reduction achievable through the application of the best system of continuous reduction which (taking into consideration the cost of achieving such emission reductions and any non-air quality health and environmental impact and energy requirements) the Administrator determines has been adequately demonstrated." The Noise Control Act of 1972 requires standards that are "feasible and protect public health and welfare, taking into account the magnitude and conditions of the use of the product, best available control technology, and costs." OSHA must issue standards that "assure so far as possible every working man and woman in the Nation safe and healthful working conditions." CPSC standards are designed to "protect the public from unreasonable risks of injury."

2. *Are performance standards required by statute?* Under the FWPCAA, EPA is required to issue performance standards; performance standards also apply to the Noise Control Act and the NSPS. Under the Occupational Safety and Health Act and the statutes administered by CPSC, the agencies must demonstrate that a significant risk to health and safety exists and that their regulation would correct it.[19] Until recently, these corrective measures could have been either performance or design standards, although design standards have dominated past regulatory activity. Since the passage of the 1981 Consumer Product Safety Amendments (P.L. 97-35, Title 12, Subtitle A), Congress now requires CPSC to issue only performance standards.

3. *Was there a single official in the agency with day-to-day responsibility for each set of rules? Was that person an expert or professional in the area being regulated?* All of the regulatory processes featured one official with day-to-day responsibilities for a set of rules. This person was typically part of the office or division responsible for developing the rules, as opposed to the division responsible for assessing economic effects. For BPT, NSPS, noise abatement, and CPSC rules, the technical project officer typically was an engineer. OSHA standards, however, were typically overseen by a project officer with an industrial hygiene

---

[19] In order to designate a product a "hazardous substance" under the Hazardous Substances Act, CPSC must first hold an adjudicatory hearing.

or toxicology background. This officer worked closely with a project task force that also included an attorney.

4. *Could the agency require firms to submit data?* The FWPCAA (section 308) and the Clean Air Act (section 114) give EPA authority to require plants to submit data or to enter plants to obtain data. For the other three rulemaking procedures, no such authority exists, and rulemaking must rely on voluntary submission of data. Whether the lack of authority to compel data submission makes the rulemaking processes at OSHA and CPSC different from those at EPA depends both on the comprehensiveness and quality of existing data in the public record and on the willingness of firms to volunteer information. These factors vary from case to case.

5. *Did contractors perform background information studies?* To a greater or lesser degree, all agencies relied on consultants and other extra-agency sources for gathering and analyzing information—for the BPT and noise emission standards, this reliance on contractors was nearly complete. Contractors made initial recommendations on what the regulations should be, and virtually all of the technical and economic information these agencies used came from their consultants. Contractors were regularly used to develop the NSPS. At OSHA, contractors wrote most of the background information studies. OSHA also depended upon its sister institution, the National Institute for Occupational Safety and Health (NIOSH), for technical information. Until recently, contractors performed most of the economic impact studies at CPSC. Some of the technical information was developed internally and some at the National Bureau of Standards. Before 1981, the CPSC had some of its standards developed externally through an "offeror" process by which the commission permitted public interest organizations and other groups to develop a proposed regulation. The Consumer Product Safety Act Amendments of 1981 closed off this avenue, except that an offeror can submit an existing or revised voluntary standard for CPSC consideration in response to an Advanced Notice of Proposed Rulemaking.

6. *Were both economic and technical analyses performed on each rule? Were different groups responsible for each type of analysis?* All of the rulemaking processes used both types of analyses for each rule. Separate divisions at EPA had responsibility for the two types of analysis of BPT rules, each group using its own contractors. Different contractors developed noise regulations also, but each contractor was responsible to the same project officer. For the NSPS rules, the same office conducted both technical and economic analyses, although usually through two contractors supervised by separate branches within the office. The Office of Policy Planning and Evaluation performed a second-stage review of

economic findings. In the late 1970s, OSHA hired economists who assumed responsibility for economic analysis; previously one group had
responsibility for both technical and economic analyses. At CPSC, different groups within the agency were assigned responsibility for technical
and economic aspects of rulemaking.

7. *Did the agency and its contractors work closely with industry to
develop standards? Did the agency rely primarily upon industry for data?*
For BPT, NSPS, and consumer product safety standards, a substantial
amount of communication took place between the agency and the industries to be regulated, because industry holds most data upon which
these rules are based. EPA used site visits to develop some BPT regulations and made extensive use of them in developing NSPS regulations.
CPSC inaugurated an experimental program to encourage industry to
develop voluntary standards in lieu of regulation—something of an
industrial counterpart to the offeror process for consumer groups. In
addition, CPSC generated much technical information independently of
affected industries. Furthermore, because all meetings between CPSC
staff and the public (except those involving proprietary information)
must be advertised and open, opportunities for ex parte contact were
minimal.

OSHA has tried to maintain an "arm's length" relationship with industry without the benefit of CPSC's requirements concerning outside
contracts. Noise rules were developed initially with little industry input.
Equipment purchasing and testing, as well as demonstration projects,
provided much of the data. The noise office relied on the comment
procedure as the primary conduit for industry data.

8. *Were "working groups" used to resolve intra-agency disputes?*[20] All
of the rulemaking processes used internal procedures to resolve disputes.
The smaller agencies, CPSC and OSHA, used informal procedures,
while EPA tended to formalize conflict resolution in bona fide working
groups.

9. *Was interagency review used?* During the Nixon administration,
OMB imposed on EPA a requirement that it submit its proposed rules
to other federal agencies for a "quality of life" review. Although CPSC
and OSHA were originally spared this requirement, OMB currently
reviews all major rules federal agencies propose.[21]

10. *Did an advisory group influence rulemaking?* Most of these rulemaking procedures included an advisory group, specified by statute, to

---

[20] These groups consist of personnel drawn from interested units in the agency, headed
by a person from the office with responsibility for issuing the regulations. The groups'
recommendations are forwarded to higher level officials for review and approval.

[21] See Executive Order 12291, *Federal Register*, February 19, 1981.

assist the agency in its deliberations. These groups did not participate with equal vigor in the rulemaking process. The extent to which they participated probably depended jointly on the agency's policy and on the advisory group itself. At one extreme, EPA's Effluent Standards and Water Quality Information Advisory Committee was ineffectual. At OSHA, an advisory committee discussed general policy issues but did not get involved in making specific rules (although OSHA occasionally established ad hoc advisory committees to help consider certain rules). In contrast, the advisory committees for NSPS and CPSC played active roles, not only in gathering information but also in making decisions. The noise control program had no advisory group.

11. *Were public comments received from all interested parties?* The BPT rulemaking procedure provoked many critical comments from industry. Comments in favor of more stringent regulation were conspicuously absent—the agency received virtually no substantive comments from environmental groups. EPA's experience with comments on the noise and NSPS standards was similar.[22] In contrast, OSHA and CPSC received comments on regulations from all interested parties in more nearly equal proportion. Labor unions and consumer groups were especially active participants in the OSHA regulatory process and the proposed CPSC regulations, respectively.

12. *Were deadlines tight?* With the exception of CPSC, each of these agencies had deadlines of some kind imposed upon them. BPT regulations for thirty industrial categories were to be issued within one year. The Clean Air Act Amendments of 1977 gave EPA only ten months to promulgate certain NSPS rules.[23] EPA had no specific timetable for producing noise emission regulations; but once the agency identified a major source of noise, it was required to promulgate a regulation for that source within two years. OSHA faced a deadline only if it issued an Emergency Temporary Standard (ETS). According to Section 6c of the Occupational Safety and Health Act, OSHA may promulgate (without rulemaking) an ETS upon evidence of an imminent and grave danger to workers. The ETS serves as a proposed permanent regulation, but it expires if OSHA does not promulgate final regulations within six months.

The lack of deadlines at CPSC and at OSHA (unless an ETS is issued)

---

[22] One important exception to this rule is the NSPS for emissions from fossil-fuel-fired utility boilers, in which environmental groups played an important role. See Ackerman and Hassler (1980).

[23] The 1977 amendments to the Clean Air Act require EPA to designate other new sources and promulgate regulations for 50 percent of them within two years, 75 percent within three years, and 100 percent within four years. See P.L. 95-95, August 7, 1977.

means that analysis had the potential to be more deliberate, with greater attention given to detail and with less propensity for error. At the same time, groups with a stake in the outcomes could have had more time to organize, analyze, and bring both political and analytical pressures to bear on agency decision makers. To the extent that differences in the outcomes of the regulatory process hinge on the existence of deadlines, the BPT case probably would not be broadly representative of processes without deadlines.

13. *Were supporting documents distributed for comment before the formal proposal appeared in the Federal Register?* EPA distributed supporting documents at the beginning of the two public comment periods. OSHA made such distributions occasionally but never for regulations that began as Emergency Temporary Standards. At CPSC, the proposed regulation appearing in the *Federal Register* was the earliest public version.

14. *Must hearings be held by statute?* Hearings are required at OSHA and CPSC. In addition, under some of the statutes CPSC administers, the hearing officer must permit some limited cross-examination of agency witnesses; otherwise, these hearings are legislative in nature, held not for the purpose of making judgments or findings but for generating information. Hearing officers may, at their discretion, permit cross-examination of some witnesses. Although EPA is not required to hold hearings, its policy is to honor all requests for hearings. Only three public hearings were held on the BPT rules.

15. *Were many rules challenged in court?* With the exception of the NSPS, a large fraction of the regulations these agencies promulgated were challenged in court. Industry typically has been concerned with the overall stringency of the regulations. Various public interest groups have also brought actions, although these suits usually concerned timing or other procedural aspects.

16. *How many rules were issued under each procedure?* Table 2-1 shows that the number of regulations promulgated as part of the BPT process exceeds that of any of the other four rulemaking processes. In fact, the table probably understates the difference in regulatory activity for at least two reasons. First, the Group I effluent standards were promulgated during a three-year period, between 1973 and 1975; the others have been promulgated during a much longer period. Second, what is counted as a BPT regulation in the table is actually a set of regulations. For the leather tanning industry, for example, EPA promulgated four separate types of regulations (BPT, BAT, NSPS, and pretreatment standards) for six subcategories, covering seven pollutants in each subcategory, with both one-day and thirty-day standards for

each pollutant. Thus, for that industry alone, EPA issued a total of 336 separate standards.

17. *Are the data documenting how the rules were set available in the public record?* The officials contacted as part of this study claimed that documents are generally available to describe the regulations and to provide the technical and economic information used to justify the regulations. Examination of EPA's BPT documentation, however, puts this claim in doubt. Further research is required to judge whether empirical study similar to that presented in this book could be supported with data available from another process. Documentation on noise regulations is available through the National Technical Information Service, OSHA documents may be found in dockets for each set of rules, BPT and NSPS documentation is in the EPA library in Washington, D.C., and CPSC documents are available at the CPSC library in Washington, D.C.

18. *Was categorization a feature of the rulemaking process? How was it used?* All of the rulemaking processes categorized regulatees into groups for the purpose of setting standards. In the BPT and NSPS procedures, subcategories within industry categories were often created for specific production processes. In addition, small plants were frequently placed in a zero-control subcategory. Regulatees of both processes pressured for additional subcategories attuned to their special circumstances. Because NSPS major industry categories were more narrowly defined than those for BPT, interested parties may have felt that further subcategorization of NSPS categories was not necessary. The relatively moderate use of categorization in the NSPS process may be explained by the use of standard-setting formulas that take industry heterogeneity into account.

Because CPSC regulations had a direct and more easily traceable effect on health and safety, economic considerations tended to be deemphasized in the categorization decision. For instance, small plants were not exempt from regulation. Rules for CB radio antennas, however, pertain only to omnidirectional, not directional, antennas because design changes in the omnidirectionals cost less and they were in wider use than directional antennas.[24] Because of the sweeping nature of some CPSC regulations (for example, those pertaining to drug packaging), the agency often granted exemptions for products that were shown to fall outside the purview of the regulation. Such exemptions may be considered categories without standards.

[24] See *Federal Register*, August 8, 1982.

OSHA also used explicit categorization less than that found in the BPT and NSPS processes. The typical OSHA health exposure standard set a maximum concentration applicable to all places of employment.[25] Categorization, if used at all, might grant different compliance dates to different industries or exempt industries with characteristics that would make certain standards inappropriate. For example, agriculture, pesticide application, and the use of arsenically treated wood are exempt from the inorganic arsenic standard, which is meant to apply to arsenic manufacturing.

Categorization was also not used extensively in regulations issued under the Noise Control Act, primarily because the regulations focused on specific products such as motorcycles, locomotives, and portable air compressors. Some categorizations based on size, age, or engineering design considerations can be found, however.[26]

## APPLICABILITY OF STATISTICAL ANALYSIS

This comparison of BPT rulemaking process with four others leads to the conclusion that documentation of the decision-making trail for the four other processes is probably sufficient to allow the use of the same procedure as for analyzing the BPT process. Moreover, sufficient numbers of regulations were promulgated in at least the NSPS and CPSC procedures to support statistical analysis.

In the crucial areas of agency discretion, bureaucratic structure, and information acquisition, the five regulatory processes are quite similar. The principal difference between the BPT process and the other four is that the BPT process produced far more rules than any of the others in a much shorter period of time. The reasons for such a difference are speculative—the BPT process may have been somehow more efficient at generating rules, the issues involved less contentious, or the deadline imposed by Congress and enforced by the Natural Resources Defense Council consent decree more effective. This difference argues for caution in drawing inferences about the entire class of technical social rulemaking from the analysis of the BPT process alone.

---

[25] The cotton dust standard is the primary exception to this general conclusion. Cotton dust exposure standards vary by the stage of cotton processing.

[26] Harley-Davidson, a major U.S. motorcycle producer, asked for relaxation of the proposed 78 dB noise standard on grounds that the design of its best-selling motorcycle would have to be drastically altered to meet the standard. At the same time, its Japanese competitors were expected to have no trouble meeting the same standard. At least in part because of the complaint, the standards for all road motorcycles were relaxed to 83 dB in the 1980 model year and to 80 dB in the 1983 model year.

# REFERENCES

Ackerman, Bruce, and William Hassler. 1980. *Clean Coal/Dirty Air* (New Haven Conn., Yale University Press).

Davis, K. C. 1978. *Administrative Law Treaties*, vol. 1 (San Diego, Calif., K. C. Davis).

Hamilton, R. W. 1972. "Procedures for the Adoption of Rules of General Applicability: The Need for Procedural Innovation in Administrative Rulemaking," *California Law Review* vol. 60, p. 1276.

Leman, Christopher. 1980. "How To Get From There to Here: The Grandfathering Effect in Public Policy," *Policy Analysis* vol. 6, no. 1, pp. 99–116.

Scalia, A. 1981. "Back to Basics: Making Law Without Rules," *Regulation* (July/August) pp. 25–28.

U.S. Congress, House of Representatives. 1977. "Clean Air Act Amendments of 1977," report no. 95-564, 95 Cong. 1 sess.

Williams, Stephen. 1975. "Hybrid Rulemaking under the Administrative Procedure Act: A Legal and Empirical Analysis,"*University of Chicago Law Review* vol. 42.

# 3

# Informal Rulemaking in Practice: BPT Standards

Translating general propositions concerning regulation into specific hypotheses about a particular rulemaking process requires detailed knowledge of both procedural and behavioral features of the particular rulemaking process. The process chosen for statistical testing is the setting of best practicable technology (BPT) standards for industrial water pollution control by the Environmental Protection Agency (EPA) during the 1972 to 1975 period. Information on this process was obtained from a series of extensive interviews with participants in the process, a review of internal and published EPA documents, and a review of several case studies of BPT rulemaking in particular industries.

The choice of the BPT rulemaking process was based on three criteria. First, this process generated enough regulations to allow statistically powerful tests of the hypotheses. Second, the process took place during a relatively short period (three years) so that the effects of changes in the economy, public opinion, or Congress are minimized. Finally, data on the rules themselves and on the factors that influenced them were accessible.

The Federal Water Pollution Control Act Amendments of 1972 (FWPCAA) redirected the national effort at controlling wastewater discharges by replacing ambient water-quality standards with a series of uniform national effluent standards for industrial point sources.[1] The

---

[1] For a history of earlier attempts at controlling water pollution, see Irwin and Selig (1975). The 1972 act also subjected publicly owned treatment works to performance standards based on secondary treatment and offered grants for the construction of treatment facilities.

first target date was July 1, 1977, when industrial point sources were to be required to meet BPT standards. By July 1, 1983, additional standards based on the application of the "best available technology economically achievable" (BAT) were to be applied. The act directed the EPA administrator to define the precise quantity of pollutants that could be discharged after the installation of these two technologies, as well as to establish new source performance standards (NSPS) for facilities yet to be constructed and pretreatment standards for facilities discharging into municipal sewage plants.

The act instructed EPA, within one year of passage, to publish NSPS guidelines for each of thirty priority industries, called Group One. The agency attempted to promulgate BPT, BAT, and pretreatment standards along with the NSPS regulations.[2] Because EPA doubted that it could promulgate standards for all activities within the Group One industries listed in the act, it divided some of them into two parts, or phases. Most large industries with high levels of water use were divided into phases. During the course of the rulemaking process, each phase was often divided into subcategories that could, in the agency's judgment, be appropriately covered by a single set of regulations.

## THE BPT RULEMAKING PROCESS

The rulemaking process can be divided into four distinct stages: (1) Contractor Study Stage—the production and distribution of the technical contractor's Development Document; (2) Proposed Rule Stage— the agency proposal of effluent standards; (3) Promulgated Rule Stage— the agency promulgation of final standards; and (4) Judicial Review Stage—litigation, settlements, and revisions. Figure 3-1 provides a schematic view of the four stages of the rulemaking process.[3]

The EPA administrator assigned responsibility for rulemaking to the Effluent Guidelines Division (EGD) of the Office of Water Planning and Standards.[4] A project officer—usually a chemical or sanitary engineer familiar with one or more industries—was responsible for shep-

---

[2] The law also directed the administrator to regulate additional industries if he determined it to be necessary. Although they were somewhat less important as sources of water pollution, an additional eighteen industries were added to the regulatory program and became known as Group Two. The rulemaking process for Group Two industries had somewhat different features and, being later in time, operated in a different political and economic setting.

[3] For additional information, see Burt (1977), Gaines (1977), and Urban Institute (1977).

[4] Appendix 3-A presents a simplified organization chart linking the various components of the agency involved in the effluent guidelines program.

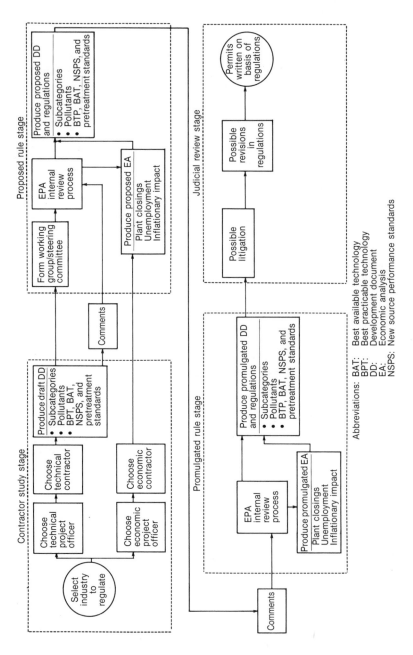

**Figure 3-1.** The industrial water pollution standard-setting process

32

herding the regulations for several industries through the regulatory process. Interviews with project officers and division managers indicated that many members of the Effluent Guidelines Division shared a strong sense of professional responsibility to identify the "best" technologies for the basis of the BPT and BAT standards. Some agency officials also voiced the need for the division to act as an agent for environmental interests, who did not participate actively in the rulemaking process for individual industries.

Because EPA felt that earlier industrial water pollution studies were inadequate, it contracted with outside consulting firms for technical studies of each industry. Each study was to be supervised by the designated EPA project officer. The contractors were to use primary sources, such as plant visits and surveys; secondary sources, such as Refuse Act Permit Program permits that predated FWPCAA; and expert opinion to analyze wastewater problems and treatment alternatives in the industry and to recommend technology-based standards.

The contractor reports were all required to have the same basic organization:

1. *A general description of the industry*. The contractor was provided some discussion of the scope of the industry—numbers and locations of plants, size of shipments, and production processes in use. Some contractors were able to offer detail on all the plants in the industry, while others presented only industrywide data that reported total production of all plants. Some discussions of the production processes were highly technical and lengthy, while others were very brief.

2. *Industry subcategorization*. The contractor also offered its judgment on whether the industry categories should be further divided into subcategories for the purpose of water pollution regulation. It was to evaluate several factors to determine if sufficient variability existed among plants to justify establishing separate subcategories. Such factors included raw materials, finished products, production processes, size and age of plants, wastewater characteristics, and treatability of wastes.[5] Raw materials inputs, finished products, and production processes were most often used as the basis of subcategorization. Size and age of plants were used less often, while differences in wastewater characteristics and treatability were judged to be adequately reflected by industry subcategories based on the other factors.

---

[5] Some contractors tried to persuade EPA to use another factor, impact on receiving-water quality. EGD, however, maintained that the FWPCAA prohibited such a consideration.

3. *Wastewater characterization.* The contractor defined the "raw waste load" in this section of the report. Based on its understanding of the processes employed, the contractor determined when water was introduced into the process and became contaminated and ascertained the volume of water and contaminants that did not return to the process and that subsequently became wastewater.[6]

4. *Pollutant parameter selection.* This section was to cover the significance of the pollutants in the industry's wastewater. The contractors regarded certain pollutants as insignificant either because their concentrations were low or because they were not harmful; conversely, the contractors selected pollutants to regulate because they appeared in large quantities or in high concentrations or because they caused harm even at low concentrations.

5. *Control and treatment technology.* The contractor reported on the performance of wastewater treatment techniques used in the industry, reviewing in-plant controls, end-of-pipe controls, pilot plants, laboratory results, and demonstration projects, as well as analytical data on the effectiveness of these techniques.

6. *Cost, energy, and non-water quality aspects.* The contractor was to provide a series of cost estimates for a hypothetical "model plant" in each subcategory and define the characteristics of this plant. Few model plants for Group One industries had the actual characteristics of any single operating plant in the industry, and few of them represented the average plant characteristics in the industry. These cost estimates provided the investment and operating costs associated with different levels of effluent discharge limitation by a single model plant, although estimates were sometimes presented for different sizes of plants in each subcategory. The reports also discussed the implications of waste-water control for air pollution problems, solid waste disposal, and energy requirements.

7. *Recommendations for BPT, BAT, NSPS, and pretreatment standards.* Finally, the contractor's report recommended a set of subcategory standards for BPT, BAT, and NSPS. BPT standards were to be the weakest of the three and NSPS the tightest.[7] The report also recommended pretreatment standards, but they were usually not given as much attention.

---

[6] Phase One studies excluded use of wastewater for cooling, boiler blowdown, and water treatment.

[7] These standards were usually written in terms of units of pollutant per unit of production (measured by input or output).

TABLE 3-1.   RECOMMENDED EFFLUENT LIMITATIONS FOR THE
POULTRY PROCESSING INDUSTRY

| Industry subcategory | BOD[a] | Total suspended solids[a] | Grease[a] | Fecal coliform[b] |
|---|---|---|---|---|
| Chickens | 0.46 | 0.62 | 0.20 | 400 |
| Turkeys | 0.39 | 0.57 | 0.14 | 400 |
| Fowl | 0.61 | 0.72 | 0.15 | 400 |
| Ducks | 0.77 | 0.90 | 0.26 | 400 |
| Further processing only | 0.30 | 0.35 | 0.10 | 400 |

[a] All standards except those for "further processing only" are measured in kg/kkg (or lb./1,000 lb.) live weight killed. Standards for further processing are measured in kg/kkg of poultry processed.
[b] Measured in units of maximum count per 100 ml.

The process of calculating the exact numerical BPT limitations began by defining the BPT technology. Next, the contractor determined performance of the technology at the "best" plants in the industry, as measured by milligrams of pollutant per liter of wastewater (mg/l). Based on survey data, it then estimated the waste-water flow at the plants practicing the best internal water conservation techniques, measuring this estimate in terms of liters of wastewater used per unit (usually a thousand kilograms) of production (l/kkg). Combining these two estimates provided a pollutant limitation expressed as a function of production (kg/kkg).[8]

Table 3-1 provides an example of the promulgated BPT guidelines for the poultry processing industry. As shown, this industry is divided into five subcategories, four of which are slaughtering plants based on the type of product (chicken, turkey, fowl, or duck) and one based on the type of process, if the plant further processes the poultry. Four pollutants are regulated, including two of the most common—total suspended solids (TSS) and biological oxygen demand (BOD).[9] Regulations for grease and fecal coliform appear in only a few industrial categories, mostly food processors.

[8] The Effluent Guidelines Division preferred the "pound of pollutant per pound of production" measure because it feared that if limitations were written in terms of mg/l, plants would simply increase their water use and meet the limitation through dilution. For industries that had no standard way of measuring physical output, the contractor recommended a new output measure. The industry for which this presented the greatest problems was electroplating. See "Metal Finishers Think EPA's Easing Up," *Iron Age*, September 6, 1976.

[9] BOD is a measure of the amount of dissolved oxygen necessary to oxidize the organic material in water to carbon dioxide and water through aerobic biochemical action.

The regulations are stated in terms of pounds of pollutant per thousand pounds of production (which is equivalent to kg/kkg). For the first four subcategories, the production measure is "live weight killed." The limitations for the "further processing only" subcategory are stated in terms of pounds per thousand pounds of finished product. For slaughtering plants that also do on-site rendering or further processing, additional adjustment factors (not shown) are used.

Most of the contractors also provided estimates of the industrywide investment and annual operating costs associated with installing the equipment required to meet the standards. This cost estimate usually assumed that all plants in the industry currently discharged raw waste.[10] One month before the technical contractor delivered the draft final report to the Effluent Guidelines Division, it was required to send the cost estimates to the contractor that was performing the economic analysis.

EPA faced a difficult task in obtaining technical and economic data from firms in order to make their discharge and cost estimates. The agency was given authority under Section 308 of the act to enter and inspect point sources and any records concerning discharges. The legislative history clearly indicates that this section was designed to aid EPA in obtaining reliable cost and waste-loading information for standard setting. Time constraints on the BPT rulemaking process often made the Section 308 approach to data collection unworkable, however, because of the legal delays available to industry.

Perhaps more important, forced data collection through Section 308 authority created or worsened an adversarial relationship between EPA and an industry.[11] Even when industry did provide EPA with all the data it requested, often the real problem was understanding what the data meant. Industry was capable of manipulating the rulemaking process by withholding data on costly, but effective, abatement technologies and by supplying excessive and confusing data. The organic chemical industry was singled out by an EPA project officer as being particularly recalcitrant in supplying data.

While the technical work was progressing, EPA held meetings with interested parties—such as firms, trade associations, environmental groups, and other government agencies—to disseminate information, to elicit views on how to define and subcategorize industries, and to

---

[10] Occasionally a contractor estimated the incremental costs associated with the regulations.

[11] As a backup to its Section 308 authority, EPA could threaten to obtain an administrative order prohibiting federal government purchases from the recalcitrant firm.

ascertain the availability of data. In only three instances, however, were hearings held.

The Economic Analysis Division of EPA's Office of Planning and Management began an effort simultaneous to that of the Effluent Guidelines Division to assess the economic impact of the Phase One regulations on each industry.[12] Because one role of the Office of Planning and Management at EPA was to review and critique agency regulations and policy, the division interpreted its role in the process to include not only identifying the potential economic impacts but also to make ". . . appropriate recommendations for alleviating those impacts."[13] These reviews—called "economic analysis reports"—were overseen by Economic Analysis Division project officers. Most of these officers had graduate-level training in business administration or economics but, as newly hired employees, had little prior experience with the industries assigned to them.

The economic analysis reports were based on a separate set of contractors' reports of economic impact (completed in six to nine months), and included financial profiles and projections of the number of plant closings, employment impacts, price changes, and other economic effects expected to occur in response to compliance with the regulations.[14] In addition, the incremental costs of the regulations were also estimated by multiplying the contractor's model plant costs by the number of plants subject to the regulations. In most cases, two contractors performed the economic studies. Time pressures required almost exclusive reliance on secondary sources and simplistic models of market behavior (such as the assumption that market demand is totally inelastic). Since a formal report was not due until effluent guidelines were proposed in the *Federal Register*, the project officer maintained close contact with the technical contractor to receive preliminary analyses as soon as possible. Often drafts of parts of the economic analysis reports were made available throughout EPA as early as the completion of the contractor's report.

The economic analysis reports and contractor reports accepted by the project officers were distributed to a wide audience, including industry and other internal and external reviewers. Usually a thirty-day comment period was allowed, although on occasion it was extended. Comments from industry were generally quite specific, relating to contractor errors,

---

[12] There were two exceptions to this statement. The economic studies for the seafoods and the rubber industries were the responsibility of the Analysis and Evaluation Division of the Office of Water Programs and Standards almost from the start.

[13] Davis (1974).

[14] Plant closings were defined as occurring whenever abatement costs were expected to exceed one-half of net profits.

assumptions, methods, and judgments; for instance, the contractor's definition of the "best" technology often came under much criticism. Reviewers also made general comments such as that abatement costs would be unreasonably high. A reading of the summaries of comments and EPA responses to them in the *Federal Register* suggests that EPA found the most persuasive arguments to be those that either identified errors in the EPA analysis or provided new data that supported alternative standards.

The Effluent Standards and Water Quality Information Advisory Committee, which was set up under the FWPCAA, routinely reviewed many of the rules, but apparently with little effect. Under its quality of life review process, the Office of Management and Budget (OMB) also screened the effluent regulations. OMB's only ability to influence the standards, however, depended upon its right to hold up the rulemaking process while it reviewed EPA regulations, a right which the Natural Resources Defense Council's (NRDC) consent decree eliminated.[15] NRDC sued EPA to force the agency to promulgate the effluent guidelines in compliance with the schedule mandated by Congress. EPA settled this suit by agreeing to a rigid timetable for issuing the standards. OMB's ability to offer critical review was limited both by time and by the formidable technical skills required.

The Commerce Department often participated in setting individual industry regulations, but it was not staffed to provide original technical responses and usually only passed on comments supplied to the department by industry. The Economic Policy Staff of the Commerce Department harshly criticized the methodology used to construct EPA's economic analysis reports, but with unknown effects.

Within EPA, the contractors' reports, written comments, preliminary Economic Analysis Division findings, and in-house expertise were the inputs to working groups headed by the Effluent Guidelines Division and including high-ranking members of other EPA offices. Consensus rules reached here were sent to the assistant administrators and, eventually, to the EPA administrator for approval. Contractors' recommendations were modified only if the Economic Guidelines Division was convinced that technical changes were warranted or if it decided that compromises were necessary to obtain approval for the rules.

The proposed rules were circulated in "development documents" (which were usually amended technical contractors' reports) and the *Federal Register*. The economic analysis reports analyzing the amended rules were also distributed. EPA then provided outsiders another thirty days

---

[15] *NRDC* 6 E.R.C. 1033.

to comment on the proposed rules; often, this comment period was extended by thirty to forty-five days through a notice in a later *Federal Register*.[16] This comment period usually resulted in more formal responses from the industry, which by this time had been given about six months to scrutinize the document that formed the basis of the proposed rules.[17] Nonetheless, most of the comments on the proposed rules were restatements of criticisms made in the earlier comment period.

Most of the comments came from individual firms and trade associations rather than state water pollution authorities, members of Congress, or environmental groups. Despite intense interest of Congress in water-quality improvement and its decision to follow the technology-based standards approach of the 1972 Water Pollution Control Act Amendments, it generally paid little attention to the details of the effluent guidelines program. The staff of the Senate Subcommittee on Air and Water Pollution felt that the program was too complicated for effective oversight, especially because the staff did not possess the technical expertise to deal with the issues. In any case, during the later years of the Nixon presidency, the staff was primarily interested in the construction grants program for municipal treatment plants rather than the industrial effluent guidelines program of the act.

Even though the technical nature of the guidelines thwarted congressional committee oversight, individual members of Congress might still have conveyed their constituents' concerns directly to EPA and attempted to inject political considerations into the deliberations. Interviews with EPA personnel at all levels, from project officer to administrator, indicated that this kind of political persuasion was usually of little import. The agency could always respond that, as much as it shares the elected representative's concerns, it had no choice because the NRDC consent decree forced it to promulgate regulations without any delay for further analysis or deliberation and because the "hard numbers" in the development documents and economic analysis reports forced it to base regulations on that "best" technology. Without a good understanding of the discretion available to EPA under the act and without the expertise to cope with the technical arguments, individual members of Congress were usually powerless to push their cases further within the agency. Nonetheless, in some instances, individual members were per-

---

[16] Rules for eight industries were proposed in December 1972. These industries complained that the Christmas mail crunch prevented them from meeting EPA's deadlines for commenting.

[17] For example, the Utility Water Act Group submitted four bound volumes of comments totaling more than 500 pages on the powerplant industry rules.

suaded to write letters to EPA on behalf of plants located in their districts.

Environmental groups participated in few, if any, of the rulemaking processes for setting industrial effluent standards. These groups appear to have made this strategic decision for at least two reasons. First, the environmentalists believed EPA, or at least the Effluent Guidelines Division, to be sufficiently committed to the goal of improving water quality that it would use its own technical expertise to support as stringent, or almost as stringent, discharge regulations as they could credibly propose, even with the assistance of outside consultants. Second, and more important, the environmental groups reached a consent decree with EPA in 1973 that forced the agency to meet a strict timetable for promulgation. As a comprehensive strategy, promulgating effluent standards for all industrial dischargers within two years probably achieved more total waste discharge reduction than if more time had been devoted to promulgating standards in each industry. More detailed analysis by EPA might have led to more stringent regulations in some industries but at the costs of significant delay in implementation and the risk of failing ever to promulgate standards for some industries with particularly complex waste treatment problems.

The second comment period was, in the agency's view, less important than the earlier one.[18] In actual practice, individual project officers were deeply involved in studying and proposing regulations for the Phase Two portions of the industry and could give little time or attention to the comments on Phase One proposed rules.

After final rules were promulgated in the *Federal Register*, legal challenges to these rules could be made within ninety days. In our sample of Group One industries, some 70 percent were active in individual or consolidated postpromulgation court actions, and more than half of these actions resulted in changes in the standards.

## CHANGES IN EFFLUENT GUIDELINES

The rulemaking process used to set effluent guidelines has changed surprisingly little since the original BPT, BAT, NSPS, and Pretreatment

---

[18] Deputy Administrator John Quarles, in a statement before the House Public Works Committee in July 1974, characterized the second comment period as a ". . . wholly one-sided stream of critical commentary on our proposals." EPA's official view was that "In most cases major issues or objections to the approach taken or the conclusions reached in the EPA draft report will have already been raised by the contractor's draft report. Criticisms of the adequacy of the data base and the analytical methods employed should therefore be expressed now rather than after the notice of proposed rulemaking." *Federal Register*, August 6, 1973.

Standards were promulgated between 1973 and 1975. The simplified version on the flow chart in figure 3-1 would look much the same today, except that OMB has a much larger role in screening regulations (as part of its mandate under Executive Order 12291), both before they are proposed and before they are promulgated.

The first major change in rulemaking was the passage of the 1977 amendments to the Federal Water Pollution Control Act, which relaxed deadlines for the BAT phase of rulemaking.[19] In this phase, the act required that standards for biological oxygen demand, total suspended solids, and other "conventional" pollutants be set according to "best conventional technology" (BCT). Thus, BAT was to apply solely to toxic pollutants—specifically the sixty-five pollutants listed in the 1977 act. Although 60 percent (eighteen out of twenty-nine) of the industry categories now have BAT standards, the practical significance of the amendments is still not clear. Many of the toxic pollutant standards may already be met through attainment of BPT standards for total suspended solids and for toxic pollutants. No studies appear to have been made of the net effects of the BAT/BPT revision. The process is currently being conducted with more deliberation, both because of the less stringent deadlines and because of Executive Order 12291, which mandated cost-benefit and economic analyses for all major rules.

Whatever its effects on regulatory outcomes, the 1977 act did not alter the regulatory process per se. Recently, however, dissatisfaction with the cumbersomeness of the process has led to internal efforts to stream-line it. These efforts include early involvement of senior staff, elimi-nation of review loops and nonessential participants, improved coor-dination with OMB, early Science Advisory Board review, and limitation of comments on proposed rules to a sixty-day period.[20] These efforts do not appear to involve fundamental changes in the regulatory process.

A number of other procedural and organizational changes have oc-curred. For example, technical contractors are given twelve months rather than six months to complete their development documents, and internal review time has also increased. In addition, because tensions developed between the Effluent Guidelines Division and the Economic Analysis Division, which between 1973 and 1975 reported to different assistant administrators, the economic analysts' work has been shifted to the Water Programs Office.[21] Now, a single assistant administrator

19 P.L. 95-217, December 27, 1977.
20 See "Final Report of Effluent Guidelines Streamlining Study," Office of Management and Evaluation, EPA, February 16, 1982.
21 The Effluent Guidelines Division reported to the Assistant Administrator for Water Programs, while the Economic Assessment Division reported to the Assistant Adminis-trator for Planning and Management.

with overall responsibility for the entire effluent guidelines program can resolve any disagreements between the technical analysts and the economic analysts.

The Effluent Guidelines Division has grown from twenty professionals in 1973 to fifty in 1983. This growth led to reorganizations that inserted another layer of management between the branch chiefs and the division director and added four more branch chiefs to the original three. Relaxing the time pressure and hiring more staff has brought the Effluent Guidelines Steering Committee back into the rule-approval process. The working groups themselves are now more organized, meet more frequently, are better attended, and tend to be less dominated by the Effluent Guidelines Division.

The division still relies heavily upon contractors for technical support, but now the development document contracts are usually bid competitively. In addition, in the original round of effluent guidelines rulemaking, the contractors not only analyzed alternative treatment technologies but also suggested specific technologies for BPT and BAT; under current agency practice, the contractors refrain from recommending technologies and standards. The Effluent Guidelines Division also holds more public hearings than it did earlier, because the Clean Water Act requires an opportunity for public hearing prior to the promulgation of pretreatment standards (a division focus since 1976). The hearings have not been well attended, however. Finally, the division has taken over the Office of Permit Programs' responsibility for sending promulgated standards to regional and state permit authorities.

Perhaps the most significant changes in the effluent standard-setting process have originated from forces at least partly outside of the agency. Both President Carter's Executive Order 12204 and President Reagan's Executive Order 12291 (which superseded E.O. 12044) directed EPA to issue more cost-effective rules.[22] In conforming to the Carter order and the intent of the 1977 Act, EPA funded studies to determine the marginal cost of secondary treatment by municipal waste treatment facilities. The resulting number was to be used as a benchmark for judging the stringency of industrial BCT regulations. If a proposed set of standards was expected to result in marginal costs exceeding the benchmark, those standards were to be relaxed.

[22] Section 3 of Executive Order 12291 requires agencies to prepare and consider a regulatory impact analysis in promulgating all major rules. This requirement exceeds the demands of Executive Order 12044 (*Federal Register*, March 24, 1978) by requiring agencies to consider the expected benefits and costs of a proposed regulation and to determine the level of "net benefits" through a benefit-cost analysis. Much uncertainty remains concerning the required thoroughness of the benefit analyses and whether, to be adopted, a major rule must yield aggregate measurable benefits that exceed costs.

OMB reviewed the results of the studies, conducted its own study with the Council on Wage and Price Stability, and concluded that the number EPA calculated as a benchmark was far too high. By this time under the Reagan Administration, EPA developed a new benchmark figure and an additional cost test. The new test has now been formally proposed.[23]

Assuming that the marginal cost tests are promulgated, the requirement that standards be cost-effective may significantly affect the process for setting effluent guidelines. Rather than calculating marginal costs, EPA contractors participating in the BPT process would typically have calculated the expected annual and investment costs for plants in the subcategory, as well as price and employment changes and plant closings expected to result from attempts to meet the standards. These other regulatory effects are only loosely related to the cost-effectiveness criterion, however, because the theoretically correct rule for setting cost-effective standards is to equate the marginal costs of abatement across subcategories. Thus, application of the marginal cost rule is likely to affect the final outcomes of BCT rulemaking and do so in a more predictable and direct way than the surrogate measures used to set BPT standards. As a result, these surrogate measures are likely to become less important determinants of the stringency of effluent standards.

Executive Order 12291 may affect both BCT and BAT rulemaking through another feature: the granting of additional power to OMB to delay or modify major regulations. OMB's role in the BPT rulemaking process under the 1972 FWPCAA was quite limited, especially once the NRDC consent decree imposed deadlines. Under Executive Order 12291, however, agencies must prepare a regulatory impact analysis (RIA) of all major rules and transmit it along with the rules to OMB for review at least sixty days prior to a proposed rule and at least thirty days prior to a final rule.[24] OMB then has up to sixty days to review the proposed RIA and up to thirty days to review the final RIA. Agencies cannot issue rules until they have responded to OMB.

The opportunities for delay are obvious and, in practice, the OMB review process has slowed down the rulemaking process. Whether the standards issued are any different from those before Executive Order 12291 is difficult to evaluate. Documentation of the decision to set a

---

[23] See Harrison and Leone (forthcoming) for a more detailed description of this episode.

[24] Executive Order 12291 defines a "major" rule as one that will likely result in an annual effect on the economy exceeding $100 million; a major increase in costs or prices; or significant adverse effects on foreign competitiveness, innovation, productivity, or employment. The EPA "Agenda of Regulations for 1981" (*Federal Register*, April 27, 1981), which listed industrial effluent guidelines under consideration, indicated that seven of these guidelines were classified as major and required a RIA.

standard is probably more thorough than it would have been without the executive order. However, the influence of OMB may be temporary. A new administration or a court ruling to speed the process, like the 1973 NRDC consent decree, could again leave OMB with little power to influence the effluent standard-setting process.

On its face, Executive Order 12291 could have profound effects on the process, both directly through OMB review and indirectly through reliance on marginal cost tests. In practice, however, the magnitude of its effects are in doubt.[25] Debate about the meaning of the executive order is still fierce, and OMB officials say they are limited to advising the agencies.

As for the marginal cost test, according to one EPA official, although BCT rules proposed before the test was created have been delayed, they are unlikely to fail the cost test. That is, the test is not going to constrain BCT outcomes. Furthermore, the marginal cost test does not apply to BAT rules.

Perhaps the greatest effect on the regulatory process is the delay arising from the executive order and the relaxed deadlines built into the 1977 clean water amendments. The delays probably result in greater attention to details, fewer errors, and a more relaxed atmosphere for conducting supporting analyses. A more relaxed atmosphere has no obvious correlation to tighter or weaker standards, however, and the process for setting BCT and BAT standards is not clearly different from that for setting the BPT standards.

[25] *Inside EPA*, December 23, 1973 (Washington, D.C., Inside Washington Publishers).

## APPENDIX 3-A
*EPA Organization Chart, 1975*

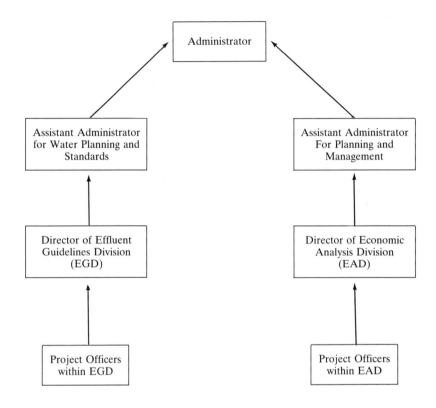

46

# REFERENCES

Ackerman, Bruce, and William Hassler. 1980. *Clean Coal/Dirty Air* (New Haven, Yale University Press).

Burt, Robert. 1977. "Effluent Limitations Under the Federal Water Pollution Control Act.," *Decision Making in the Environmental Protection Agency* vol. 2A (Washington, D.C., National Academy of Sciences).

Davis, Swep. 1974. "Methodology for Assessing the Economic Impact of Water Pollution Controls on Industrial Discharge," in *Water-1974 I. Industrial Waste Treatment*, symposium series no. 144 (New York, American Institute of Chemical Engineers).

EPA, Office of Management and Evaluation. 1982. "Final Report of Effluent Guidelines Streamlining Study," February 16.

Gaines, Sanford. 1977. "Decision Making Procedures at the Environmental Protection Agency," *Iowa Law Review* vol. 62, no. 3.

Harrison, David, Jr., and Robert A. Leone. Forthcoming. *Federal Water Pollution Control Policy* (Washington, D.C., American Enterprise Institute).

Irwin, William, and Edward Selig. 1975. *Enforcement of Federal and State Water Pollution Controls: A Report to the National Commission on Water Quality by the Environmental Law Institute* (July).

Urban Institute. 1977. "EPA's Development of Effluent Guidelines for the Beet Sugar Processing Industry," in *Organization Analysis of the Regulatory Process: A Comparative Study of the Decision Making Process in the Federal Communications Commission and the Environmental Protection Agency*, prepared for NSF Grant No. APR 75-16718 (Washington, D.C., Urban Institute, November 30).

# 4
# Rulemaking Theories and Hypotheses

This chapter presents two types of complementary models about agency behavior. One focuses on the reaction of government decision makers to perceived or anticipated signals from external groups; the other addresses the internal organization and information flow within the agency and between the agency and outside parties. These models become the basis for hypotheses about how certain factors—such as compliance costs, trade association activities, and participation in rulemaking by submitting formal comments—affect the stringency of the rules a regulatory agency sets. These hypotheses, in turn, are useful in their own right as mechanisms for developing and organizing an understanding of technical rulemaking by all social regulatory agencies. They are particularly important as a guide for the empirical analysis of regulatory behavior.

## THEORETICAL ANTECEDENTS

Most of the literature on regulation addresses what Fiorina (1981, p. 7) labels "regulatory origin," as opposed to the "regulatory process." Theories of regulatory origin attempt to explain why agencies make specific decisions such as to regulate (or deregulate) some industries or economic activities but not others or to promulgate and enforce specific regulations that harm some groups while benefiting others. Examples of theories based on regulatory origin include the public interest theory (that reg-

48

RULES IN THE MAKING

ulators seek to further the broad interest of the general public)[1] and the capture theory (that industries and occupations "capture" regulators to gain a legal cartel manager, the agency).[2] In contrast, theories of the regulatory process seek to explain how regulatory agencies make decisions, in other words, what factors determine or explain their decisions. The latter literature is more directly relevant to building models and generating hypotheses by inferring from the rules themselves the factors that influenced EPA's effluent guidelines.

The questions of how and why regulations are issued are related. The implied decision rules for EPA may explain the trade-offs the agency makes in issuing regulations, but they do not explain why particular factors influenced the agency's decisions and why they were given particular weights in reaching those decisions. Opposing theories of regulatory origin provide radically different interpretations of the decision rules. The agency-dominance view that characterizes some streams of political science (for example, Wilson, 1974), as well as the concern in administrative law with the problem of unbridled agency discretion (for example, Stewart, 1975), hold that regulatory agencies have considerable freedom in selecting their own rules for making decisions. Under this view, empirical results based on decisions by an agency reveal only that agency's own rules for setting standards.

In contrast, the congressional dominance theory (for example, Weingast and Moran, 1983, and Weingast, 1984) argues that agencies serve Congress well or, more accurately, that they serve the wishes of congressional committees. According to this theory, active congressional involvement with regulatory agencies is rarely needed precisely because Congress has created an incentive system for guiding agency decisions that is usually strong enough to make more direct measures, such as oversight hearings, unnecessary. The congressional committee system matches each representative with committees that can best promote the interests of his or her own constituency, often at the expense of general societal interests. Agencies compete with each other for a larger share of budget appropriations from these powerful committees in return for

[1] See Mitnick (1980) and Levine (1981) for recent articulations of the public interest theory. The "expertise model" of administrative law (Stewart, 1975) takes a somewhat similar view that regulators should be insulated from political pressures in order to allow them to channel their expertise into the service of public interest objectives well-defined by Congress.

[2] Stigler (1971) provided the first formal statement of the capture theory based on the "rational actor" view of economics. Kolko (1965) offered an early explanation for the capture of a commission, and Peltzman (1976) provided an important extension of the Stigler approach.

servicing committee members' constituencies (Fiorina and Noll, 1978). In addition, agency administrators and commissioners must be confirmed by the Senate.

The central conclusion of the congressional dominance theory implies a quite different interpretation about regulatory agency decision making. From this viewpoint, the regulations an agency promulgates and the decision rules its actions imply would reveal congressional preferences rather than agencies' unbridled discretion. Thus, agency decision rules would reflect congressional interests, which implies that regulatory reformers must focus their attention on changes within the structure of Congress if they want to change the rulemaking process.

*External Signals Theory*

In Noll's (1976) version of external signals, groups external to an agency measure its success and provide favorable or unfavorable signals both inside and outside agency proceedings. Because the definition of success varies from one group to another, the agency seeks regulatory outcomes that balance conflicting demands, "minimizing conflict and criticism which appear as 'signals' from the economic and social environment from which they operate" (Mitnick, 1980, p. 138).

In Noll's external signals model, the courts uphold appeals of decisions that are fairly and nonarbitrarily reached; Congress approves budgets for and holds noncritical oversight hearings on agencies that perform to their liking; the Office of Management and Budget (OMB) acts favorably on budget requests of successful agencies; the press provides favorable coverage to agencies that perform well; and constituents, including those representing the beneficiaries and targets of regulations, refrain from complaining to Congress, the executive branch, or the press about successful agencies. According to this theory, the agency attempts to obtain approval from all of these groups or at least to avoid negative feedback from them.

Noll's theory suggests that agencies develop complicated administrative processes to protect (for example, by limiting the number of organized groups that can afford to participate in the process) groups that approve their actions and to assess the positions of the interest groups involved. More important, the theory suggests that agencies will respond to all groups that can provide positive rewards or negative sanctions to the agency, as opposed to being captured by one or a few private special interests or by a crusade to serve the public interest.

Stewart's "interest representation model" (Stewart, 1975) views agency proceedings as a political arena in which competing groups urge their own views upon the agency. This model reasons that wide participation in the rulemaking process (and in judicial review) leads to agency decisions that serve the public interest, just as supposedly occurs in the legislative process. As Stewart so forcefully argues, however, granting open access to the administrative process does not guarantee active participation by all affected parties or necessarily better agency decisions.

As a normative solution to the agency discretion problem, the interest-representation model may fail, as Stewart explains. Nonetheless, as a positive model of rulemaking under hybrid rulemaking procedures, this model provides important insights into the decision-making process and closely parallels the Noll model of agency responses to external signals.[3]

## Information-Based Theories

The models in this study[4] primarily rely upon the external signals theory. These externally focused models should be supplemented, however, by hypotheses generated from models of organizational structure and information flow. The external signals theory ignores organizational factors that may influence the decisions an agency makes. Debate continues on organizational questions such as whether agencies should be run by administrators or commissioners and whether they should be independent of the executive office.[5] These internal factors are arguably important determinants of agency behavior, and they should be systematically examined, just as are the external influences on agency behavior.[6] With

[3] The bargaining theories of Dahl and Lindblom (1953), Holden (1966), and Wilson (1974) can be interpreted as a variant of the external signals theory. In these theories, the agency uses a bargaining process to balance the signals received from the regulated industries, consumer interest groups, and other interested government agencies. Some bargaining theories differ from the external signals theory in that the proponents of bargaining theories are reluctant to give up a role for the public interest in the bargaining, or balancing, process. Dahl and Lindblom, for example, argue that commissions enter into this bargaining process because they lack power to enforce unilateral decisions and that they represent the bargaining position of the "unrepresented" public interest.

[4] The study describes the setting of Best Practicable Technology industrial effluent guidelines by the Environmental Protection Agency.

[5] See, for example, the U.S. President's Advisory Council on Executive Organization (1971), Noll (1971), and Committee on Government Affairs (1977). The recent controversy over the management of EPA's toxic waste program has resurrected many of these issues.

[6] Roberts and Bluhm's (1981) book is an excellent example of this type of research into internal influences on an organization's decisions. They provide case studies of the internal structure of decision making within six electric utilities and then attempt to derive general conclusions about the effects of variations in internal structure on the behavior of these firms.

some exceptions (primarily for internal information flows), however, this study does not explicitly consider internal factors.

The complementarity between external signals and internal information flow can easily be seen. Participants in the rulemaking process influence agency decisions through the signals they provide. These external signals convey information about the amount and distribution of costs and benefits of proposed agency actions, as well as about the participants' preferences. Agencies seek information from external sources through the rulemaking process because they lack sufficient information about costs, benefits, and preferences—or more precisely, because some of this information is costly to acquire. Indeed, regulation can be viewed as an information-producing process.

Information from external sources has value to the extent that it is unavailable within the agency, influences agency decisions, or reduces uncertainty about the agency's own information. Thus, the internal bureaucratic factors that determine the amount, type, and credibility of information generated within an agency also can significantly affect the agency's regulatory decisions.

*Bounded Rationality Theories.*   Economists and decision theorists typically assume comprehensive rationality, that is, that decision makers can generate all possible alternatives to a problem, assess the probabilities of all consequences of each alternative, and evaluate each set of consequences on the basis of their own goals. In technical rulemaking, however, such an assumption clearly seems unwarranted; the problems are too complex and the available time is too short. The bounded rationality approach (Simon, 1957; March and Simon, 1958) more closely matches this type of rulemaking. It postulates that limited time and limited information-processing and problem-solving ability force decision makers to adopt shortcuts. For example, because problems are so complex, they must be factored into subproblems that can be solved more or less independently by different parts of the organization. Also, because of the constraints placed upon the decision makers they forgo the goal-optimizing approach for one that meets all goals "well enough" (the "satisficing" approach). Of importance for our empirical study is the hypothesis that most organizations face problems similar to those already solved, so they develop standard operating procedures to limit the extent of choice in repetitive problems. This shortcut implies that regulatory agencies facing complex, but structurally similar, decisions develop rules for making such decisions.

As further developed by Cyert and March (1963), this theory stresses that organizations learn over time, suggesting that a regulatory agency adapts its standard operating procedures in reaction to its past successes and failures and in reaction to changes in the agency's external environment (such as change in presidential administration).[7]

In the policy analysis literature, Porter and Sagansky (1976) recognize that the creation, transfer, and withholding of information are important determinants of regulatory agency decisions. Although they do not refer to the bounded rationality approach, they suggest three hypotheses that follow from this theory: (1) the economic efficiency of regulatory decisions is determined by the available expertise, (2) an agency's decisions are constrained by practical and administrative problems of gathering and analyzing data, and (3) the regulated firms affect decisions by the key information that they do (or do not) provide. Porter and Sagansky's fourth hypothesis—that agency decisions attempt to balance political forces acting on it—essentially restates the external signals theory.

Downs (1967) offers a long list of hypotheses to assist in predicting the behavior of government agencies. His approach appears quite similar to that of Cyert and March. Both suggest that organizations will attempt to reduce uncertainty and will concentrate on a short-run approach to problems. Down's hypotheses are not very useful in structuring an empirical study, however, because they require comparing one agency's rulemaking decision rules with those of another agency or with the same agency's rules on different dates. Data for such comparisons are not generally available. One exception is Downs' "law of increasing conservatism," which suggests that, all other factors being equal, rules promulgated early in a rulemaking process will be more strict than those promulgated later.

*Theories from Information Economics.*    Models that explain the role of information in the regulatory process are not yet well developed in economics. Most of the economics literature on information theory focuses on market structure and organizational issues within profit-seeking firms rather than government agencies, with the following exception.[8] One of the greatest problems for agencies performing rate-of-return regulation is that they cannot acquire accurate information about the future costs of providing service. Recognition of this problem has led to work on the design of regulatory systems that induce regulated firms to supply accurate cost information for ratemaking.[9] The role of infor-

[7] See Moe's (1982) analysis of how presidential party changes affect regulatory decisions.
[8] See Williamson (1964), Marshak and Radnor (1972), Alchian and Demsetz (1972), Spence (1975), Mirrlees (1976), Stiglitz (1979), and Beales, Craswell, and Salop (1981).
[9] See Loeb and Magat (1979), Sappington (1983), Vogelsang and Finsinger (1979), and Williamson (1976).

mation has received much less attention in the literature on social regulation. Obtaining accurate cost information is still a problem, but it occurs in the context of designing regulatory incentives to induce regulated parties to reveal their expected compliance costs.[10]

The economics literature on information theory, though usually expressed in terms of the theory of the firm, implies several hypotheses about the effect of information on agency rulemaking decisions. First, the greater the quantity of information an agency has about the firms it regulates, the greater its ability to pursue its own goals without bowing to external pressures. For example, the more the Consumer Product Safety Commission (CPSC) knows about alternative mechanisms for making a household product safe, the lower the probability that an intervenor will suggest a new protective measure during hearings and that a reviewing court will remand the standard to CPSC for further study of alternatives to its proposed standard.

Second, higher quality information in support of a proposed regulation reduces opponents' ability to modify the regulation, either by presenting conflicting information during the rulemaking process or upon judicial review. Again, the agency can more easily pursue its own goals rather than those of the participants.

Third, the sequence in which information is received may affect its influence on rulemaking decisions. Ingram and Ullery (1977) suggest that early information has greater effect than information provided later, because once agencies develop opinions about what is reasonable regulation, those opinions are usually difficult to alter.

## THE STRINGENCY OF REGULATIONS—AN EXTERNAL SIGNALS VIEW

How can the stringency of regulations across different parties be compared? The answer requires finding a meaningful metric for making comparisons. For EPA's industrial effluent discharge standards, for example, one possible measure of stringency is the percentage of all pollutant discharge that each plant must eliminate. The Occupational Safety and Health Administration (OSHA) standards for different hazardous substances, usually written in units such as pounds of substance per cubic yard of interior air, can be measured by the percentage reduction in accidents or deaths caused by exposure to the substance or the number of accidents or deaths per exposed worker

---

[10] See Baron (1984), Groves and Loeb (1976), and Sonstelie and Portney (1983).

per year. For CPSC, design standards for regulated products, such as swimming pool slides, lawn mowers, cribs, and match covers, can be translated into units of accidents or deaths per exposure day or into units of percentage reduction in accidents or deaths.[11] Because CPSC uses hospital emergency room accident data as one basis for selecting the order in which products should be regulated, these alternative measures are not as unusual as they might first appear. More will be said about the measurement question in the next chapter. For now, stringency may retain its somewhat imprecise meaning without impeding the theoretical analysis in this chapter.

According to the external signals theory, a regulatory agency sets the stringency of its standards to maximize the net support for a regulation that it receives or expects to receive from interested outside parties, subject to a constraint on total agency resources. The regulation is assumed, in effect, to transfer wealth from firms, which oppose more regulation, to citizens, who support it. This assumption simplifies the regulatory process by requiring that only two groups compete against each other.[12]

In measuring the influence of external signals on agency decisions, it is important to use a broad definition of "influence," as suggested by Nagel (1975) and Arnold (1979). Their definition identifies two possible causal links between the preferences of external parties and an agency's decisions: the direct link between actions taken (that is, signals sent) and agency behavior, and the indirect link through which the agency makes decisions in anticipation of the likely signals its actions would elicit from external parties. To the extent that an agency is successful in reducing direct adverse signals by anticipating the external reactions

[11] Of course, these transformations themselves can be quite controversial because often the factors necessary to make the transformations are not known with certainty.

[12] The assumption does seem generally appropriate for social regulation, which usually seeks to curtail a previously permissible practice in the interest of a broad social objective. For economic regulation, however, support or opposition cannot be so easily characterized. Some firms may reasonably view regulation as a useful anticompetitive device to help in imposing cartelization, setting minimum prices, preventing entry, or reaching some other anticompetitive objective. Even for social regulation, the assumption that industry opposes and the public supports more regulation is often questionable. For example, a proposal by the CPSC to mandate safety clutches on new lawn mowers may be opposed by consumers less concerned about safety than about the impact of such devices on the price of the product. Likewise, a proposed EPA regulation that affects only a portion of an industry is likely to receive support from unaffected firms interested in improving their market positions, although this support is likely to be overshadowed by opposition from that portion of the industry whose relative position is worsened. Appendix 4-A on industry categorization addresses this issue of interfirm rivalry created by regulation.

to unfavorable decisions, it is influenced more by the second causal link than by the first one.[13] Thus, an external signals model must include both measures of direct signals and measures of factors that influence agency actions through anticipated reactions.[14]

The process of anticipating external signals may work through an internal agency debate in which agency employees act as agents of outside interests. For example, the interviews carried out in connection with this study reveal that the business-school-trained analysts in the Economic Analysis Branch of the EPA water office were more attuned to the unfavorable cost impacts on firms than were the engineers in that office's Effluent Guidelines Division. Some engineers felt that part of their role was to act as agents for environmental interests. As this example makes clear, some signals that influence regulatory agency decisions are generated internally. Nevertheless, for the sake of consistency, the "external signals" label will continue to be used. The external signals are distinguished from internal ones by their origination in sources other than bureaucratic organization or information flows.

## The Stringency Decision Model

Peltzman's (1976) model of government behavior in response to competing private interests offers a convenient framework for making operational the notion that a regulatory agency sets the stringency of its standards to maximize net outside support for a regulation, as perceived or anticipated by the agency. In Peltzman's stylized version, citizens vote either for regulators or for elected officials who directly control regulatory decisions (the congressional dominance hypothesis); an agency makes decisions to maximize favorable votes. Although using the same basic structure, this study's model has a different focus from Peltzman's. To recognize the possibility that congressional interests and regulatory decisions may be incongruent, our model assumes that the agency max-

---

[13] Weingast and Moran (1983) use this same distinction between directly observed behavior (in their case, associated with congressional control over a regulatory agency) that influences agency actions and agency actions associated with its anticipation of congressional reactions. The addition of this second cause of agency actions explains the seeming paradox of minimal congressional oversight.

[14] The perception of the importance of various signals may be agency-specific and may, in fact, vary over time within the same agency. For example, EPA under the Reagan administration probably treats complaints from environmentalists quite differently than it did under President Carter. Accordingly, the external signals approach will be most useful in explaining differences in regulatory outcomes at a specific agency within a limited time period, unless the influence of the change in administrations can be easily captured, say, by using a dummy variable.

imizes the net level of external signals it expects to receive. Two groups primarily determine the net level of external signals $M$: the $n$ "citizens" who benefit from the cleaner homes, workplaces, or environment the agency's regulatory actions produce; and the $m$ "firms" that create the regulated hazards. The questions of whether these signals are sent directly to the agency or indirectly through Congress is open.

For a given regulation, the agency does not know with certainty how any one citizen or firm will respond to a particular standard or how the responses will differ among citizens and among firms. The agency can, nonetheless, assign a probability $f$ that a representative citizen will offer a positive external signal and a probability $h$ that a representative firm will oppose the regulation, that is, convey a negative external signal.[15] The factor $e$ measures the relative signal strength of the representative firm and citizen—that is, $e$ conveys the relative ability of the two types of signals to influence the agency's decisions. Thus, its value depends upon the preferences of the agency for pursuing "firm" values versus "citizen" values and upon the anticipated actions firms and citizens are likely to take to influence the decision-making process.

Both probability functions, $f$ and $h$, depend upon the stringency $s$ of the regulations the agency sets. For environmental regulation, $s$ is defined to be the standard limiting the total amount of pollution discharge by the entire industry. Industry standards can take any nonnegative value, with $s = 0$ corresponding to a complete elimination of the hazard and higher values of $s$ corresponding to less stringent standards. The probability functions, $f$ and $h$, depend upon the agency's perceptions of the citizens' benefits and firms' costs from regulation. These probabilities can take on any value between 0 and 1 for they are the agency's perceptions of the proportions of citizens and firms who will participate in the regulatory process. Not all will participate.[16] Because of transaction and information costs, individual citizens and firms may have only imperfect information and may refrain from expressing their preferences about a regulatory decision. Also, individual preferences vary among

---

[15] The use of a "representative" firm or citizen is an obvious oversimplification, because the diversity of interests within an industry or citizenry may have real effects on regulatory outcomes. This construct will be relaxed in appendix 4-A on categorization. Although some citizens may oppose some proposed agency actions, even when compared to the status quo, the assumption that the "representative" citizen supports agencies seems reasonable. Of course, the intensity of that support depends upon the benefits the proposed actions offer the representative citizen. Similarly, some firms may actually support some proposed regulatory actions, while the "representative" firm opposes those actions. The strength of its opposition depends upon the estimated costs of the actions.

[16] See Peltzman (1976) for further justification of this assumption.

both citizens and firms, even given the same standard and its associated benefits and costs. Thus, averaged across citizens and firms, some of whom will participate in the regulatory process and some of whom will not, the probabilities $f$ and $h$ must take on intermediate values.

The agency's assessed probability of citizen support $f$ depends upon the stringency of the industry standard $s$ through the intermediate variable $v(s)$, the total damage that each citizen suffers from the activities of all regulated firms in the industry. Note that $s$ is an industry standard that corresponds to the aggregate impact on all the firms the agency regulates. This formulation implies a total limitation of $s$ on hazardous exposure from all firms (such as an EPA limitation on discharges from all sources of organic wastes entering a lake). In this example, $v(s)$ could be the adverse health effects upon each individual from an EPA ambient water pollution standard of $s$. As this example illustrates, $s$ influences public damages, such as water or air pollution, to each citizen independently of the damage any other citizen incurs. Less stringent standards (that is, higher values of $s$ that imply more discharge) are assumed to increase damage $v(s)$ at an increasing rate (that is, $v'(s) > 0$ and $v''(s) > 0$).[17] Higher damage also is assumed to reduce the perceived probability of citizen support at an increasing rate (that is, $f'(v) < 0$ and $f''(v) < 0$).[18]

Define $t(s/m)$ as the cost to a representative firm of an industry standard set at level $s$ when $m$ firms are in the industry. The argument of this firm's compliance cost function is $s/m$, the average firm discharge. Compliance cost $t(s/m)$ is measured by the loss of profits that meeting the standard would cause. This cost incorporates the expenses of purchasing and operating control equipment; it also reflects the extent to which these costs can be passed on to consumers in the form of higher prices (and to taxpayers in the form of tax credits). It is through this influence on profits that regulations induce opposition from firms.[19]

The agency's assessment of the probability of a firm's opposition $h$ depends upon the firm's compliance costs $t$, which in turn depends upon

---

[17] In this book, the prime and double prime superscripts indicate the first and second derivatives, respectively, of a function.

[18] The assumptions that $v''(s) > 0$ and $f''(v) < 0$ are sufficient to guarantee the concavity of the objective function (4.1). The assumption that $v''(s) > 0$ is consistent with most of the theoretical and empirical literature in environmental economics. The assumption that $f''(v) < 0$ is based on the intuition that victims of increasingly severe pollution damage become increasingly vocal in their opposition within the political process that determines the level of damage.

[19] The consequences of incurring compliance costs—such as unemployment and plant closings—influence the stringency of agency standards by affecting the relative signal strength $e$.

the stringency of the industry standard $s$. More stringent standards (lower values of $s$) are assumed to impose higher costs at an increasing rate (that is, $t'(s/m) < 0$ and $t''(s/m) > 0$). Also, higher costs are assumed to raise the agency's perceived probability of firm opposition at an increasing rate (that is, $h'(t) > 0$ and $h''(t) > 0$).[20]

Combining the strength of anticipated signals from supporting citizens and opposing firms gives the (net) external signals level,

$$M = nf[v(s)] - meh[t(s/m)] \tag{4.1}$$

The agency selects the stringency of the standard $s$ to maximize $M$ as defined in equation (4.1). Thus, the politically optimal standard satisfies,

$$nf'[v(s^*)]v'(s^*) = eh'[t(s^*/m)]t'(s^*/m) \tag{4.2}$$

where $s^*$ is the value of the standard at the maximum value of $M$. The agency sets $s^*$ to equate the sum of its anticipated political support and the sum of its anticipated political opposition at the margin. Figure 4-1 summarizes the stringency decision model.

In contrast to the politically optimal standard $s^*$, the economically efficient level of the standard $\hat{s}$ occurs when the marginal cost of increasing the stringency of the standard (reducing $s$) equals the marginal benefit of that increase in stringency; that is, when

$$-t'(\hat{s}/m) = nv'(\hat{s}) \tag{4.3}$$

A comparison of equations (4.2) and (4.3) shows that the rulemaking process leads to the socially efficient standard if and only if

$$-f'[v(s^*)] = eh'[t(s^*/m)] \tag{4.4}$$

Equation (4.4) requires that at the politically optimal level of the standard $s^*$, a dollar increase in compliance cost to a firm raises its anticipated strength of opposition (weighted by its relative signal strength) by the same amount that a dollar decrease in damages to a citizen raises that

---

[20] Consistent with footnote 18, we assume that $h''(t) > 0$, because firms become increasingly active in their opposition to higher compliance costs as those costs escalate. To guarantee interior solutions, which we assume in the comparative statics analysis, two further assumptions are necessary. One relates to the curvatures of the citizen support function $f$ and the firm opposition function $h$, and the other to the requirement that these probabilities be between 0 and 1. First, damage $v$ cannot exceed the value at which $f(\bar{v}) = 0$, and second, compliance costs $t$ cannot exceed the value at which $h(\bar{t}) = 1$.

*Objective function:*

$$M = n\ f[v(s)]\ -\ me\ h[t(s/m)]$$

*Variable definitions:*

$M$ = net level of external signals anticipated or perceived by the agency
$n$ = number of citizens
$s$ = aggregated industry discharge standard
$m$ = number of firms
$e$ = relative signal strength of representative firm and citizen

*Function definitions:*

$v(\cdot)$ = total damage to a representative citizen
$t(\cdot)$ = compliance cost of a representative firm
$f(\cdot)$ = agency's perceived probability of citizen support
$h(\cdot)$ = agency's perceived probability of firm opposition

**Figure 4-1.**   The stringency decision model

person's anticipated support for the agency. The politically optimal standard, $s^*$, does not necessarily satisfy equation (4.4), however. In other words, no *self-regulating* mechanism exists to guarantee that the political choice of the standard is economically efficient.

Figure 4-2 illustrates graphically how $s^*$ and $\hat{s}$ are determined. The relationship between the level of the economically efficient standard $\hat{s}$ and that of the politically optimal standard $s^*$ depends upon the values of the parameter $e$ and upon the relative heights of the functions $h'[t(s/m)]$ and $f'[v(s)]$. The function $h'[t(s/m)]$ measures the change in the probability of opposition from a firm produced by a dollar increase in average compliance cost, that is, the sensitivity of the firm to variations in cost. The function $f'[v(s)]$ measures the change in the probability of support from a citizen with a dollar increase in the damage from the regulated hazard, that is, sensitivity of citizens to a change in damages.

From figure 4-2, for an industry which consists of firms that are both sensitive to compliance costs (high $h'$) and are politically powerful (high $e$) and which faces little sensitivity to damages ($|f'|$ low), this term $|f'[v(s^*)]|$ would be less than $|\langle eh'[t(s^*/m)]|$. From equation (4.2) this configuration implies that marginal damages are greater than marginal costs, or that $|v'(s^*)| > |t'(s^*/m)|$. Thus, the agency's politically chosen standard would be less stringent (have a higher value of $s$) than the efficient level of the standard. This is the situation illustrated in figure 4-2. Similarly, for industries that are not sensitive to compliance costs, are politically weak, and face a highly sensitized public, the standards would tend to be overly tight.

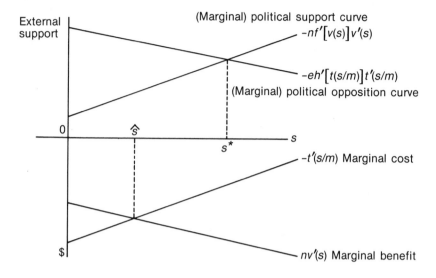

**Figure 4-2.** Comparison of economically efficient and politically optimal standards

In summary, in this theory, three key variables affect whether an industry's standard is too strict or too lenient: (1) the public's sensitivity to damages, (2) the firms' sensitivity to compliance cost increases, and (3) the relative signal strength of the firms and citizens.[21]

*Implications of the Stringency Decision Model*

Beyond its use for comparing the politically optimal standard $s^*$ with the economically efficient standard $\hat{s}$, the stringency decision model yields several interesting hypotheses about the determinants of $s^*$, many of which are tested in chapter 6.

One useful way to derive hypotheses about the external factors affecting the stringency of standards is to relate the factors to parameters of the model. From the first-order condition (4.2), we observe that the value of $s^*$ depends upon the values of three parameters, $n$, $m$, and $e$, as well as upon shifts in the six functions, $f'(\cdot)$, $v(\cdot)$, $v'(\cdot)$, $h'(\cdot)$, $t(\cdot)$, and $t'(\cdot)$. By inserting multiplicative shift parameters, $a$, $b$, $r$, and $p$, before the $h$, $f$, $t$, and $v$ functions, respectively, shifts in all six of the functions can be characterized by changes in these four parameters. With this formulation, the agency's perceived probability of a firm's

---

[21] The relationship also depends upon the parameter $m$, which will be discussed in the next section.

opposition becomes $a(h)$; the agency's expected probability of a citizen's support equals $b(f)$; the cost of complying with a standard equals $r(t)$; and the damage associated with a standard equals $p(v)$.[22]

The technique of comparative statics analysis[23] can now be used to relate changes in the optimal standard $s^*$ to changes in the following seven parameters:

$n$ = the number of citizens affected by the regulation

$m$ = the number of firms affected by the regulation

$e$ = the relative signal strength of representative firms and citizens

$a$ = the shift parameter on the agency's perception of the probability of a firm's opposition $h(\cdot)$

$b$ = the shift parameter on the agency's perception of the probability of a citizen's support $f(\cdot)$

$r$ = the shift parameter on the compliance cost function $t(\cdot)$

$p$ = the shift parameter on the damage function $v(\cdot)$

After incorporating the four parameters $a$, $b$, $r$, and $p$ into the objective function, equation (4.1) becomes

$$M = nbf[pv(s)] - meah[rt(s/m)] \tag{4.5}$$

The corresponding first-order condition for $s^*$ then becomes

$$nbf'[pv(s^*)]pv'(s^*) = eah'[rt(s^*/m)]rt'(s^*/m) \tag{4.6}$$

The comparative statics analysis proceeds as follows. Rewrite equation (4.6) as

$$
\begin{aligned}
0 = &\; Z\,(s, n, m, e, a, b, r, p) \\
= &\; n \cdot b \cdot f'\,[p \cdot v(s)] \cdot p \cdot v'\,(s) \\
&- e \cdot a \cdot h'\,[r \cdot t(s/m)] \cdot r \cdot t'\,(s/m)
\end{aligned} \tag{4.7}
$$

suppressing the superscript on $s$. The total differential of (4.7) is

$$
\begin{aligned}
0 = &\; \frac{\partial Z}{\partial s}\,ds + \frac{\partial Z}{\partial n}\,dn + \frac{\partial Z}{\partial m}\,dm + \frac{\partial Z}{\partial e}\,de \\
&+ \frac{\partial Z}{\partial a}\,da + \frac{\partial Z}{\partial b}\,db + \frac{\partial Z}{\partial r}\,dr + \frac{\partial Z}{\partial p}\,dp
\end{aligned}
$$

[22] The multiplicative form in which the shift parameters $a$, $b$, $r$, and $p$ are modeled is not the only way to represent a shift in the relevant functions. The multiplicative form is a particularly simple one that models the shifts in a reasonably neutral form. It also lends itself to unambiguous comparative statics analyses that are easy to interpret.

[23] The comparative statics analysis technique is described in many advanced microeconomics texts such as Henderson and Quandt (1971).

where

$$\frac{\partial Z}{\partial s} = (nbp)\,(f'v'' + v'^2\,f''p)$$
$$- (ear)\,(h't'' + t'^2h''(r/m)) < 0 \tag{4.8}$$

$$\frac{\partial Z}{\partial n} = bf'pv' < 0 \tag{4.9}$$

$$\frac{\partial Z}{\partial m} = (ears)\,(1/m^2)\,(h''rt'^2 + h't'') > 0 \tag{4.10}$$

$$\frac{\partial Z}{\partial e} = -ah'rt' > 0 \tag{4.11}$$

$$\frac{\partial Z}{\partial a} = -eh'rt' > 0 \tag{4.12}$$

$$\frac{\partial Z}{\partial b} = nf'pv' < 0 \tag{4.13}$$

$$\frac{\partial Z}{\partial r} = -(eat')\,(h' + rh''t/m) > 0 \tag{4.14}$$

$$\frac{\partial Z}{\partial p} = nbf'v' < 0 \tag{4.15}$$

From (4.7) through (4.15),

$$\frac{\partial s}{\partial n} = -\frac{\partial Z}{\partial n}\bigg/\frac{\partial Z}{\partial s} < 0$$

$$\frac{\partial s}{\partial m} = -\frac{\partial Z}{\partial m}\bigg/\frac{\partial Z}{\partial s} > 0$$

$$\frac{\partial s}{\partial e} = -\frac{\partial Z}{\partial e}\bigg/\frac{\partial Z}{\partial s} > 0$$

$$\frac{\partial s}{\partial a} = -\frac{\partial Z}{\partial a}\bigg/\frac{\partial Z}{\partial s} > 0$$

$$\frac{\partial s}{\partial b} = -\frac{\partial Z}{\partial b}\bigg/\frac{\partial Z}{\partial s} < 0$$

$$\frac{\partial s}{\partial r} = -\frac{\partial Z}{\partial r}\bigg/\frac{\partial Z}{\partial s} > 0$$

$$\frac{\partial s}{\partial p} = -\frac{\partial Z}{\partial p}\bigg/\frac{\partial Z}{\partial s} < 0$$

The effects of parameter changes on the stringency of the standards $s^*$ that are shown in table 4-1.[24] Thus, ceteris paribus, a regulatory agency

[24] For example, $\frac{2S}{\partial n} < 0$, then table 4-1 lists $m$ with a minus sign..

promulgates more stringent standards (lower values of $s$) the greater the number of citizens it protects ($n$), the higher the agency's perceived probability function for citizen support ($b$), and the higher the damage function ($p$). The agency sets less stringent standards (higher values of $s$) the larger the number of firms it regulates ($m$), the greater the relative signal strength of firms in comparison to citizens ($e$), the higher the agency's perceived probability function for firm opposition ($a$), and the higher the compliance cost function ($r$).

The results in table 4-1 allow development of potentially measurable characteristics of each industry and its regulatory environment, which are represented by the seven parameters, and formulation of testable hypotheses that link these characteristics and the parameters to the agency's decisions about the stringency of its regulations.

Only some of the potential hypotheses were subjected to statistical tests. Others were not tested because data were unavailable or because no reasonable method existed for quantifying the causal variable. A hypothesis also may be inapplicable to the specific rulemaking process examined. In the descriptions that follow, the tested hypotheses are in italics.

*Parameter n (Number of Citizens).* From table 4-1, an increase in $n$ tends to raise the level of the standards regulatory agencies set; that is, the more the potential exposure to a regulated product, the more stringent the regulation imposed upon it. A hazardous regulated substance produced in highly populated regions or in plants with many workers may cause more widespread damage than it would in less populated regions and plants. Graphically, the marginal political support curve in figure 4-2 shifts upwards with an increase in $n$. In the example of environmental regulation, regulatory agencies facing a larger number of citizens $n$ receive more individual requests for protection. More impor-

TABLE 4-1. COMPARATIVE STATICS EFFECTS ON
THE STRINGENCY OF REGULATIONS

| Parameter increased | Sign of associated change in $s^*$ |
|---|---|
| $n$ (number of citizens) | − |
| $m$ (number of firms) | + |
| $e$ (relative signal strength) | + |
| $a$ (probability of firm opposition) | + |
| $b$ (probability of citizen support) | − |
| $r$ (compliance cost) | + |
| $p$ (hazard damage) | − |

tant, citizen and labor groups have more active participants, represent
a larger number of voters, raise more money to support their activities,
and therefore may more effectively argue their interests in the regulatory
arena. Thus, those firms operating in more populated regions or selling
to a larger number of customers tend to receive more stringent standards
than other firms to the extent that the regulatory agency can associate
its standards with the affected populations.

*Parameter m (Number of Firms).*   Table 4-1 shows that the greater
the number of firms a standard affects, the less stringent the total discharge standard imposed upon the entire industry. An increase in $m$
causes an upward shift in the marginal political opposition curve, yielding
an equilibrium value of $s$ corresponding to a less stringent standard (see
figure 4-2). Although this result about the effect of the number of firms
on the total discharges allowed under the industry standards is hardly
surprising, the more interesting question concerns the stringency of
standards issued for each firm, that is, $s/m$. In figure 4-2, the outward
shift in the (marginal) political opposition curve produces a larger op-
timum value of $s$ at which the value of (marginal) political opposition,
$eh'[t(s/m)]\, t'(s/m)$, exceeds its previous value. Because this function is
inversely related to the average firm standard, $s/m$,[25] this firm standard
must be lower (that is, more stringent).
*Hypothesis 1: Industries with more firms are permitted to create more
total damage, but the average firm standards are more stringent.*

*Parameter e (Relative Signal Strength).*   The parameter $e$ measures
the relative signal strength (or political clout) of a firm as compared to
a citizen. A higher value of $e$ shifts the marginal political opposition
curve up, resulting in a weaker standard (that is, a higher $s$) (see table
4-1). Five factors could increase the value of $e$.

1. Firms employing large numbers of workers tend to have higher
relative signal strength (and therefore weaker standards) because regulatory costs imposed upon them create adverse effects for more voters.
The larger the number of employees (voters) affected, the more sensitive
agencies will be—through congressional oversight and White House
controls—to the political consequences of their actions. These employers are able to influence more voters and so are afforded more influence
in the rulemaking process.
2. Firms also weaken their standards by strengthening their political
clout through enlisting the support of Congress, the Office of Manage-

---

[25] Differentiation of $eh'[t(s/m)]t'(s/m)$ with respect to $(s/m)$ establishes this inverse relationship.

ment and Budget (OMB), the Commerce Department, and other government agencies. Individual senators and representatives appreciate the opportunity to further their constitutents' interests through telephone calls and letters to, and conferences with, the agency administrator. Members of Congress can also influence agency actions through oversight and budgetary hearings. Such actions build personal political support without appearing to compromise the interests of other constituents.[26]

The geographical location of firms may be an important determinant of their influence on members of Congress. Where one firm or many firms in the same industry dominate an area's employment or payrolls, the politicians from that area may be motivated to seek relief for those firms.

Consider the role of OMB as a regulatory facilitator. In 1971, OMB declared its intention to review agency decisions through its Quality of Life Review; Executive Order 12291 (February 17, 1981) provided OMB much stronger powers to review, delay, and even force substantive revisions in regulation. By catching the ear of OMB, firms may be able to enhance their clout in the rulemaking process. The Commerce Department, in its role as agent for business interests, also meets with regulatory agencies to discuss the interests of affected industries. However, some government bodies such as the Justice Department are available to increase the political clout of citizen interests as well (that is, to lower $e$).

3. Informal rulemaking often provides a period before the proposal of regulations for firms to comment on the standards suggested in the notice of proposed rulemaking published in the *Federal Register* or in contractor documents, as well as a period before promulgation of the final rules to comment on the proposed regulations. These comment periods provide the most important formal channel for firms to influence the regulatory process. By providing well-reasoned arguments in favor of weaker standards and supporting this position with technical arguments and data, firms can force agencies to consider their interests seriously. Many agencies include summaries of the commentors' arguments and agency responses as standard practice, and these arguments become part of the written record reviewed by the courts.[27]

---

[26] See Fiorina and Noll (1978) for a detailed version of this facilitation theory.

[27] Characterizing industry comments as a means of increasing political clout is a simplification. Many comments are likely to contain information useful to the agency in setting standards. Because firms are unlikely to provide data supportive of tighter standards, the better the quality of their comments, the more likely the standards will be relaxed.

The political salience of firms' arguments to weaken their standards is likely to be much higher if they can appeal to adverse effects of regulations on the rest of the economy, beyond the compliance costs imposed upon themselves. Besides increasing compliance costs to firms, stricter standards can cause prices to rise if the added costs can be passed on to consumers; they also may lead to unemployment and plant closings if production levels are reduced. These effects are often viewed to be important in the regulatory process. For example, in its economic analyses for air and water pollution standards, EPA regularly analyzes the extent to which stronger standards will result in higher prices, unemployment, and plant closings.

*Hypothesis 2: Active participation in the rulemaking process by firms through the submission of well-documented comments to the regulatory agency in support of weaker standards may increase their signal strength, resulting in weaker standards for them.*

4. Unemployment and plant closings create adverse community impacts that, in turn, create an unfavorable political climate for the agency, and price increases caused by regulations conflict with the important national goal of reducing inflation. At least the first two of these factors tend to sensitize regulatory agencies, Congress, and other political institutions to the arguments made by the affected firms and to provide a public interest argument for the agencies to use in support of weaker regulations. However, the fact that price increases may be viewed as under the control of the affected firms may make this variable an ineffective element in an industry's public relations arsenal.

*Hypothesis 3: In an industry category or subcategory for which these three economic impacts (the level of inflation, employment losses, and plant closures) are predicted to be high, the agency would expect stronger opposition from the affected firms, raising e and, from table 4-1, leading to a weakening of the standards.*[28]

5. The strength of opposition from firms subject to a regulation may depend upon the degree of homogeneity of that opposition within an industry category. If some firms have already installed control technology capable of meeting a set of proposed standards, then the very existence of these complying firms weakens the argument that noncomplying firms cannot overcome the technological or cost hurdles necessary to meet the standards.

---

[28] The expected compliance cost of a regulation may also affect its political import, but this variable has its own term $r$ in the model. Defining such costs as a factor affecting parameter $e$ would not affect our results. In addition, plant closures and unemployment express the consequences of incurring compliance costs.

*Hypothesis 4: The smaller the percentage of industry in prior compliance with a regulation, the more seriously a regulatory agency will consider the industry's opposition to regulation and therefore the weaker the standards it will receive.*

*Parameter a (Probability of Firm Opposition).* The product of the parameter $a$ times $h(\cdot)$ lies in the unit interval, because $ah[t(s/m)]$ is the probability, as perceived by the regulatory agency, that a representative firm will oppose its standards. For any level of compliance costs $t$, the greater the value of $a$, the greater the agency's anticipated likelihood that firms will actively oppose their standard. From table 4-1, firms more likely to oppose standards will tend to be granted weaker standards. Given the same level of compliance costs, what factors might cause some firms to be more likely than others to oppose agency regulation? One possible factor is profitability; however, its effect upon the stringency of a regulation is indeterminant. On the one hand, the agency may anticipate that profitable firms will be less likely to resist its regulations because they are less threatened, and so the agency will impose stricter standards upon them. On the other hand, the agency may anticipate more resistance from profitable firms because even if their effort at opposition is unsuccessful, the firms' profits are high enough to cover the legal, technical, and managerial resources necessary to oppose the regulation.[29] These firms are more likely to oppose regulations and, if successful, would be given weaker standards. A priori reasoning does not help resolve which of these two possible effects dominates.

The level of transactions costs is another possible determinant of the agency's perceived likelihood of opposition. Given the level of compliance costs associated with a standard, the higher the costs of organizing opposition to the standard, both internally and in cooperation with other firms, the less effective the opposition the firms can muster. Therefore, with higher organizing costs, an agency would expect less opposition from firms to the standard (and therefore impose a more stringent standard upon them). The per-firm costs of organizing an effective, common opposition to an industry standard depend directly upon the number of firms in the industry and inversely upon the commonality of their interests in resisting the standard. Ease of organization may be greatest in the more homogeneous, concentrated industries. This suggests an

---

[29] Firms failing to earn any profit lose their income tax subsidy for expenditures on regulatory compliance, so the after-tax cost of compliance is higher for them than for profit-making firms.

industry's concentration ratio as one, albeit imperfect, proxy for the level of organization costs necessary to implement effective opposition. *Hypothesis 5: Firms in more concentrated industries will receive less stringent standards than those in less concentrated industries.*

As another proxy, a strong, politically active trade association could also be expected to have lower costs of organizing to oppose a new regulation, because the existing lobbying machinery should be in place and already functioning effectively in the political arena. In addition, trade associations regularly monitor the events in the external environment of member firms and are experienced in mobilizing the firms to fight threatening regulations during the rulemaking process.

*Hypothesis 6: Firms belonging to large or active trade associations should tend to receive weaker regulations.*[30]

The degree of import competition U.S. firms experience is also a factor that may determine the probability, as perceived by an agency, that firms will oppose regulatory actions. Here an interesting distinction can be drawn between product regulation and process regulation. When industrial processes are regulated, as under OSHA standards or EPA air and water pollution regulations, foreign manufacturers are necessarily exempt. This exemption can give imports a competitive edge. Therefore, domestic opposition to industrial process regulation is likely to be directly related to import penetration. In contrast, imports are subject to product regulation, such as safety and air pollution regulations for automobiles. If anything, such regulations penalize imports. If only the U.S. market is regulated, foreign producers that compete in many countries may not be able to enjoy scale economies that are available to their U.S. counterparts. Hence, U.S. producers may actually seek health or safety regulation as a substitute for tariffs.[31]

Even though the expected compliance cost is the major component of an agency's anticipated probability of opposition function, that probability function may be also influenced by media attention focused on compliance costs required of the industry. As one example, interviews indicated that the media coverage of the Mahoning Valley steel plants' poor financial condition in the early 1970s contributed to EPA's doubts that these plants could afford the abatement costs of the proposed water pollution standards applicable to them. This example demonstrates the

---

[30] These two proxies are imperfect because they are usually presented in an aggregate form that buries information on the heterogeneity of firm interests. The degree of heterogeneity of interests more closely parallels organizational costs.

[31] This argument implicitly assumes that the gains to the American firms are not outweighed by the increased costs such a regulation would impose on them in their export role.

potential effect of media attention on increasing (that is, weakening) standards. Media attention, such as articles in trade journals, that focuses the agency's attention on the regulatory costs imposed upon firms would increase $a$ and thus tend to cause the agency to set weaker standards.

*Parameter b (Probability of Citizen Support).* Like $a$, the product of the parameter $b$ times $f(\cdot)$ lies in the unit interval, because $bf[v(s)]$ is the probability, as perceived by the agency, of support from a citizen, given the damages $v(s)$ produced under standard $s$. This probability varies across industries because citizens in areas affected by different firms may react differently to similar levels of hazards and because the agency's expectations about these reactions can be influenced by media attention.

Given the same level of damages $v(s)$, what factors would cause the agency to expect some people to offer more support for the regulatory action than others, and therefore (from table 4-1) cause the agency to set more stringent (that is, lower) standards? All else being equal, firms that create hazards in a region of the country whose population is highly conscious of hazards in the environment, workplace, or home are likely to face more stringent standards than firms located primarily in regions without this strong interest. People with this strong consciousness form groups to translate this interest into specific proposals for regulatory action, to develop background studies to support the proposals, and to participate in the rulemaking process to convince the agency to adopt their proposals. The Natural Resources Defense Council and the Sierra Club provide these services in attempting to influence EPA regulatory policy, while labor unions, such as the Oil, Chemical, and Atomic Workers Union, seek to direct OSHA policy. Media attention that increases the agency's perception of citizen consciousness also raises $b$ and thus would tend to cause it to select stricter standards.

*Parameter r (Compliance Cost).* The parameter $r$ measures the relative height of the total compliance cost curve. From table 4-1, a regulatory agency selects weaker standards for those firms facing higher costs because they are more likely to oppose the regulations.

Measuring the relative heights of total compliance cost curves across firms is difficult in practice. Using econometric analysis requires extensive data, which are often unavailable, and engineering approaches are expensive, time-consuming, and often inaccurate. At least three variables, which are directly related to the parameter $r$, are easier to specify.

1. The more a firm must spend in capital and operating costs to comply with proposed regulations, the more likely an agency will expect it to take positive action, either individually or with other affected firms, to actively oppose these regulations by participating in the rulemaking process; thus the agency responds with weaker standards.[32]

*Hypothesis 7: Higher annual compliance cost per plant or higher invest-ment cost per plant at some fixed level of compliance (for example, 85 percent reduction in pollution discharge) should result in weaker stand-ards.*

2. Small plants and older plants tend to have less efficient technologies for compliance, as well as to suffer from diseconomies of scale in com-plying with regulations.

*Hypothesis 8: Small or old plants are more likely to receive weaker stand-ards than large or new plants.*

3. The magnitude of the hazards being regulated are often directly related to the height of the compliance cost function. For example, a given workplace standard limiting the concentration of airborne asbestos is likely to be more costly to meet when uncontrolled airborne asbestos emissions per pound of product are high than when they are low. To take another example, the abatement technology used to meet water pollution standards written in terms of pounds of allowed discharge per pound of product is likely to be more expensive when raw discharges or water flow are larger at any given level of control.

*Hypothesis 9: Firms with higher emissions (before control) are likely to receive less stringent standards because their compliance costs are higher at any level of control.*

*Parameter p (Damage).* The parameter $p$ is an index of the level of ambient, workplace, or home damages from exposure to a given hazard. Variations in $p$ across firms could reflect differences in the assimilative capacity of the receiving media (such as wind speed for air pollution or age of family members for product safety). Table 4-1 shows that an increase in $p$ leads to a more stringent standard. Thus, firms that create hazards leading to relatively high damage levels will tend to receive more stringent standards. For example, such factors as high tempera-tures and low flow rates of receiving waters reduce waste-assimilative

---

[32] Parameter $r$ is a characteristic of the compliance cost function. It does not measure the level of compliance cost at any given level of the standard. Graphically, the compliance cost curves of industries with higher values of $r$ lie above the curves of industries with lower values of $r$. Thus, we implicitly assume that the compliance cost curves do not cross each other.

capacity, raising the damage curves and leading to tighter water pollution standards.


## THE CATEGORIZATION DECISION

Although the stringency decision model suffices for generating hypotheses about why regulatory agencies set more stringent standards for some firms than for others, it fails to consider a second important decision most social regulatory agencies make, that of categorization. These two decisions constitute the primary outcomes of the rulemaking process. Appendix 4-A extends the stringency decision model to consider how agencies simultaneously decide how to group firms into categories and how to determine the stringency of the standards assigned to each category of firms. In this more general model (to be called the general model), the stringency hypotheses derived from it remain unchanged from those developed in the last section and thus need not be discussed again.


## INTERNAL FLOW OF INFORMATION

In general, external factors should be more important than internal factors in explaining the variation in the stringency of regulations set by one agency that uses the same procedure for many industry categories because many of the internal factors do not vary across categories. However, one particular set of internal bureaucratic variables—those that determine the amount, type, and credibility of information generated within an agency—are possible determinants of variation in the stringency of regulations because they *do* vary across industry categories.

Our review of the information-based theories suggests hypotheses about several characteristics of regulatory agencies that can affect the stringency of the regulations they promulgate. All of the hypotheses rely on Simon's (1957) initial insight that information is costly to obtain and to process, which in turn implies both that information is a valuable commodity and that decisions will always be made on the basis of imperfect or incomplete information.

The way the agency obtains information is important. A social regulatory agency developing technical regulations generally must regulate so many sectors of the economy that it cannot become expert in any one of them. Agency officials must rely on information from outsiders, most of whom have an interest in the outcome of the regulatory process.

Furthermore, the nature of the rules themselves often induces asymmetry in the information resources of the various interested outside parties. This information asymmetry ordinarily works to the benefit of the side possessing more information, because it can control, to some extent, the flow of information. For regulations affecting specific products or industrial processes, firms within the regulated industry possess most of the important information, a situation implying that those firms dominate the external flow of information to the agency.

Of course, rulemaking agencies are more than merely passive receptors of information. If agency personnel have prior expectations or preferences for certain outcomes, they can help fulfill these expectations through their definitions of critical terms in the enabling statute and their decisions on which issues are important, what kinds of information to gather, and how this information is to be obtained and processed. Even seemingly technical details, such as the definition of test protocols or the design of sampling procedures, provide the agency with opportunities to bias the process towards desired outcomes.

Outcomes can be tilted even by neutral regulators. The effects of many decisions are so subtle that agency personnel may simply miss their side effects and unknowingly bias the process. For example, in sampling to determine the "best practicable technology" under the Federal Water Pollution Control Act Amendments of 1972, project officers or their contractors occasionally missed segments of industries by following particular sampling strategies. Unless subsequently corrected, these omission errors resulted in inappropriate standards for the plants in the unsampled industry segments. Because these plants always sought (and received) redress if their inappropriate standards were too stringent—but not if they were inappropriately weak—poor sampling design could have biased the rulemaking process toward weaker standards unless citizens were just as watchful and knowledgeable.

One value of accurate and complete information to the agency is that it makes both the proposed and final regulations more defensible and therefore less likely to be influenced by arguments to change them. In addition, parties that might consider arguing for changes in rules would be less likely to do so if the agency possesses higher quality information with which to rebut their arguments. Thus, better documentation and technical support of regulations mean less change in rules over the course of the rulemaking process and also less chance of reversal or remand if they are challenged in court. The direction of the effect depends on which groups exercise the most influence on the process.

*Hypothesis 10: If the influence of those regulated predominates and they argue for weaker regulations, then better technical support provided by the agency or opposing factions will result in less weakening of the rules.*

Access to information is important, but so is the ability to process it. Therefore, with given levels of staff and resources, the more time an agency has to process information the more likely it is to produce more stringent regulations, because they can be better supported by the available data. Conversely, tight deadlines can result in poorly supported regulations. Of course, time also works to industry's advantage in that long deadlines allow it to build a better case for weakening the standards.

Staff turnover midway through the development of a regulation is equivalent to an earlier deadline, in that lost institutional memory must be regained if the previous level of support is to be maintained.

*Hypothesis 11: Turnover of important personnel during the rulemaking process results in weaker rules when pressure from industry predominates and tighter rules when pressure from beneficiaries predominates.*

The phenomenon of organizational learning may also influence the outcomes of regulatory processes. In a succession of regulatory processes, both regulators and the regulated learn from the experience of earlier procedures. Hence, the order in which regulations are promulgated can affect outcomes. Later regulations should be more defensible because the agency has learned which kinds of documentation are likely to be needed to defend its rules successfully in court. Under the assumption that stringent rules are more difficult to defend in court than weak ones, this argument implies that regulations become more stringent over time as the agency learns more about how to justify its regulations. Meanwhile, those being regulated are also learning, although the amount of learning (and thus the ability to influence the stringency of regulations) might be limited if interindustry communication is lacking.

*Hypothesis 12: Within a set of regulations issued under the same process, earlier rules will be less stringent than later ones if, as probably occurs, the agency learns faster than the regulated firms.*[33]

Temporal considerations shift the focus of modeling away from the static to a dynamic context. Because informal rulemaking under the Administrative Procedures Act requires an agency to give notice and take comments on preliminary rules, which are then finalized, at least two sets of standards—one before and one after the comment period— will be issued as part of any procedure of this type. What is needed is a theory linking the standards set in the first stage of a rulemaking process to those set in the second and subsequent stages.

Unfortunately, the literature that analyzes the determinants of rules in a dynamic context is sparse; however, the writings of Simon (1957)

---

[33] Earlier in the chapter, we described Downs' "law of increasing agency conservatism," which argued that later rules are more stringent than earlier ones, in contrast to the hypothesis derived from consideration of organizational learning.

and Cyert and March (1963) are applicable. They argue that, because obtaining and processing information are costly, organizations devise ways to make decisions based on imperfect or incomplete information. In particular, an organization of even modest size is usually divided into several subunits, each charged with the responsibility for one aspect of the organization's overall function. Necessarily, these subunits have different missions. Cyert and March describe the potential for conflict within the organization due to these differing missions. They suggest that the organization attends to first one and then another objective, rather than optimizing all objectives simultaneously.

Consistent with this theory, rulemaking agencies are also composed of several groups. The group responsible for drafting the rules may be more concerned with maximizing the benefits of the regulation to the public. A separate group within the agency may focus upon the costs and the economic impacts of the regulation. As a consequence, the information this latter group produces can either help convince the drafting group of the appropriateness of its regulations or provide an argument for less stringent regulations, with their associated lower benefits. Conflict between the drafters of the rule and the economic analysis group could produce, over the course of a rulemaking procedure, the oscillating behavior Cyert and March described.

*Hypothesis 13: If a proposed rule is made especially stringent during an early stage of the rulemaking process, it is likely to become less stringent at a later stage.*

## APPENDIX 4-A
*The Categorization Decision: Models and Hypotheses*

Most social regulatory agencies produce rules that apply to classes of similar firms (often called categories or subcategories), rather than promulgating separate standards for each firm. Informal rulemaking was developed to a large extent to reduce the administrative costs associated with case-by-case regulatory decisions required by either formal rulemaking or adjudication. Because of requirements in the enabling statutes, case law, or recent executive orders such as 12044 and 12291, most agencies must consider all reasonable alternatives and their corresponding costs for every major regulation they set. For example, when setting air and water pollution standards, EPA hires contractors to assess the alternative technologies and their economic impacts for every class of firms it regulates. Also, in the past, CPSC used the offeror system described in chapter 2 to conduct regulatory analyses. Thus, the more finely an industry is divided into categories, the higher the administrative costs required to collect, process, and disseminate technical and economic information and to force the regulations through all the steps in the rulemaking process.

In order to reduce their administrative costs, agencies may seek to combine firms into uniform categories within which all firms face the same standards. Pursuit of the goal of (net) external signals maximization pushes the agencies in the opposite direction, however, that of designing regulations to fit the differing political consequences of regulating each individual firm. By creating separate categories for particularly recalcitrant firms, an agency can relax its standards for those firms and mitigate their opposition without weakening standards across the board. In addition, dividing firms in an industry into separate categories creates roadblocks to the participation both of those being regulated and third-party interest groups, because protest about one regulation can affect only a limited number of regulated firms. Thus, categorization can be used to narrow the scope and even the quantity of criticism from all outside groups.

For example, the enormous number of categories created by EPA for industrial effluent guidelines may have contributed to limiting criticism from environmental groups. One of the striking features of that rulemaking process was the virtual absence of their participation until the court review stage. These environmental groups certainly had the expertise to participate. They may have decided that spending resources on a specific regulation for a specific category would not have been cost-effective for them, given their limited resources and their other competing objectives. In fact, the major environmentalist action concerning the effluent guidelines program was a Natural Resources Defense Council procedural suit involving many industries that sought to force EPA to adhere to statutory deadlines.[34]

Creation of many categories by an agency can also improve the chances that more regulations will emerge from the courts unscathed. By proposing a single regulation covering all industries, or even one industry, an agency could face tremendous court battles over the final standard, risking delay and compromise. A portfolio approach spreads risks over many regulations by ensuring that court decisions and technical and economic arguments pertaining to one industrial category could not easily be applied to others.[35]

It might be argued that, given the technical basis of regulations, the categorization decision reflects engineering rather than economic considerations. In reality, however, almost every plant, if given the opportunity, could claim some special circumstances justifying a unique standard. Furthermore, agencies do invoke partly economic reasons for creating subcategories, such as those based on plant size, age, and location. These factors do not appear to receive equal consideration in categorizing every industry. Thus, much agency discretion appears to exist in allocating firms to separate industry categories, each with different standards.

## The General Model: A Verbal Description

Dividing firms into categories and subcategories raises the administrative costs of agency rulemaking; however, it also may increase the net level of external support or signals for particular rules. Thus, upon adding the categorization decision to the problem, the agency's objective becomes one of maximizing the net level of external signals, *less* the administrative costs of creating multiple categories for firms. Consider an industry with $m$ firms. The categorization decision involves selecting a

---

[34] *NRDC v. EPA*, 8 E.R.C. 1983. For further discussion of this suit and resulting consent decree, see chapter 3.

[35] The desire to obtain larger budgets or other results of congressional and public approval may also contribute to the proliferation of categories if it is thought that issuing more regulations that cover additional categories makes the agency look more productive. The emphasis placed by EPA's associate administrator on the larger number of effluent guidelines issued may illustrate this type of thinking. See the statement by John Quarles in "Implementation of the Federal Water Pollution Control Act," Hearings before the Subcommittee on Investigation and Review, House Committee on Public Works, June 25, 1974.

number of categories (less than or equal to *m*) and assigning firms to them. The stringency decision requires setting a separate standard for each category of firms. Given a set of category standards and an allocation of firms to categories, the damage to each citizen (and therefore the probability of supporting the agency) depends upon the total amount of hazard allowable under the standard system. For a regulatory problem involving pollution, the damage would be a function of the total discharges of all pollution sources.

The probability that any particular firm will oppose the agency's regulatory actions depends on the costs imposed upon that firm. These costs—or more generally, this reduction in profit—in turn depend on the firm's category assignment and the stringency of that category's standards. In addition, to the extent that compliance costs differ across firms, the standards assigned to other firms in the industry may also affect the firm's profitability by altering the competitive cost structure of the industry.

Given this more general statement of the agency's problem, how does the agency assign firms to categories? The necessary condition for an optimal assignment requires that for any firm assigned to one category, assigning that firm to any other category would reduce (increase) support from citizens more (less) than it would reduce (increase) opposition from firms, after adjusting for the difference in administrative costs due to the reassignment. How does the agency set the stringency of the category standards? Just as in the stringency decision model of the previous section, the agency must select standard levels that equate at the margin the perceived increase in citizen support with the perceived decrease in firm support.[36]

Let us explore the implications of these two rules for making category assignments and setting category standards. Tightening a category standard affects the probability of opposition from firms in two ways: *directly*, by raising the compliance costs of firms in that category; and *indirectly*, by making these firms less competitive with similar firms assigned to other categories. These firms in other categories benefit either by gaining market share at the expense of firms in the category whose compliance costs have been raised or by following increases in prices by the controlled firms by price increases of their own.

---

[36] Even though the rulemaking process does not strive to equalize the marginal cost of compliance across all firms, if our theory is correct, high compliance-cost firms would tend to face weaker standards than those able to reduce discharges at a lower cost. This argument assumes that firms cannot pass on all of the higher costs to consumers in increased prices, which is usually the case in the short run.

For industrial categories in which the competitiveness of rivals in other categories is not augmented by tighter standards (that is, the magnitude of the indirect effect is near zero), the optimal decision rule poses a relatively straightforward trade-off for the regulatory agency. In considering the tightening of a category standard, the agency must balance the increase in citizen support with the increase in opposition from firms in the category that face higher compliance costs. But if the competitiveness of firms in other categories is increased by a tighter category standard (that is, the indirect effect is positive), then the added opposition to the agency because of higher compliance costs is mitigated, or perhaps even outweighed, by the support from firms in other categories reaping a competitive advantage.

Two examples of the latter case of interfirm rivalry illustrate this point. EPA sets new source performance standards (NSPS) for both air and water pollution emitted from either new or expanded plants. Significantly, the NSPS are considerably more stringent than the standards imposed upon existing sources of pollution.The NSPS only minimally affect firms that are planning little future expansion, but they create a profitable barrier to the entry of other firms. If our theory is correct, EPA will oblige as long as the new sources are either small in number or politically weak. In contrast, the usual explanation for the fact that NSPS are more stringent than standards on existing sources is that new sources can comply more cheaply because they are not constrained by an existing stock of "dirty" capital. Our model suggests an alternative explanation based on the relative bargaining, or signaling, power of existing versus new sources of pollution.

The second example concerns existing firms that may be willing to bear the added compliance costs of more stringent standards as long as the tighter standards impose even higher costs upon competing firms. Effluent guidelines issued for the corn wet milling industry category were based on an abatement technology that was already in place, or at least inexpensive to adopt, for four out of the five direct-discharging firms in the industry. Thus, the standard required major expenditures by one firm; in other words, that one affected firm was in effect in a separate category. That single firm objected strenuously to the standard, while the other firms remained relatively passive.[37] Our model shows that the distribution of abatement costs across firms as well as the mag-

---

[37] The opposite experience occurred in the pulp and paper industry. To the astonishment of other firms in the industry, St. Regis cited the abatement performance of its exemplary plants, providing support for the proposed effluent guidelines for the industry. Evidently the regulations would not have required much compliance for St. Regis because its plants showed a close similarity to EPA's model plant on which the rules were based. Thus, St. Regis stood to gain a competitive advantage over other firms in its industry.

nitude of those costs are important determinants of industry opposition and therefore of the stringency of standards.[38]

Our model also explains why a politically sensitive regulatory agency would not necessarily pursue compliance cost minimization as a goal. When controlling a heterogeneous industry, the particular categorization strategy that minimizes costs would require the tightest controls for those firms with the lowest costs of compliance. Yet an alternative distribution of standards over categories involving somewhat smaller costs for the firms with cheaper controls (and somewhat higher costs to those with expensive controls) may result in a net decrease in opposition because of characteristics that vary systematically over the groups. Minimum opposition could occur, for instance, when the entire industry was being treated uniformly, even though total costs would not be minimized. Put another way, it is possible that a 1 percent weakening in the total industry standard could actually increase opposition to the agency if it were accomplished by giving some categories tighter standards while standards for other categories were weakened.

## The General Model: A Formal Statement

The literature contains few other efforts at formally modeling the rule-making process. Thus, this model may be of particular interest to some researchers. In addition, it can be used to design simulations that could confirm the validity of the hypotheses generated from the simpler models in the text and to generate more complex hypotheses about how agencies simultaneously select the stringency of standards and the assignment of firms to categories.

For a given standard, the agency does not know with certainty how any of the $n$ citizens or any of the $m$ firms will respond to it. The agency can, nonetheless, assign a probability $f_q$ that the $q$th citizen will grant external support (or give a positive signal) and a probability $h_j$ that the $j$th firm will oppose the standard. The agency then seeks to maximize the expected (net) level of external signals $M$, where

$$M = \sum_{q=1}^{n} f_q - \sum_{j=1}^{m} e_j h_j - C \qquad (4A.1)$$

The factor $e_j$ is the signal strength of the $j$th firm; it is a weighting factor that measures the relative strength of the $j$th firm's signal compared to that of each citizen, where each citizen is assumed to have equal influence. As an example, the employment level in the $j$th firm would be

[38] See equations (4A.8) and (4A.9).

80                                                           RULES IN THE MAKING

highly correlated with $e_j$. The probability functions $f_q$ and $h_j$ depend on
the agency's perception of the citizens' losses and firms' costs, respectively. Because of transactions and information costs, most of the individual $f_q$ and $h_j$ probabilities are greater than 0 and less than 1. More
important, the agency's imperfect knowledge of the true probabilities
make it reasonable to assume probabilities within the unit interval. For
simplicity, the administrative cost function associated with partitioning
firms into categories is assumed to be directly proportional to the number
of categories $k$, that is, $C = cK$ (where $c > 0$).

The probability of agency opposition by a firm $j$, $h_j(t_j)$, depends upon
the effect of the agency's standards on the firm's profits, where $t_j(s^1, \ldots,$
$s^j, \ldots, s^m)$ is the increase in costs (or profit reduction) under the
standards relative to the no-regulation state. In general, the firm's profits
could be affected not only by its own standard $s^j$, but also by those of
competing firms. A firm's standard $s^j$ is measured in units of hazardous
substance exposure per unit of time.

With $m$ firms, the agency can decide to divide the industry into at
most $m$ categories, although given the cost of categorization it will generally assign firms to a number of subcategories $k$ less than $m$. Let $s_i$ be
the standard set for subcategory $i$, which is often measured in units such
as pounds of hazardous material per pound of product manufactured.
Then, if firm $j$ with production level $p_j$ pounds of product per day is
assigned to category $i$, its standard is $s^j = s_i p_j$, measured in pounds of
hazardous material per day. The two primary decisions an agency makes
in its rulemaking process are (1) the levels of the standards $s_i$ for each
category; and (2) the assignments of firms to categories.[39] We specify
those assignments by the binary variable $\delta_{ij}$, which equals 1 if firm $j$ is
assigned to category $i$ and equals 0 otherwise.[40] Thus, for each firm $j$

$$s^j = \sum_i \delta_{ij} s_i p_j \tag{4A.2}$$

[39] Often a third important decision made by the agency is what substances to regulate
for each category of firms. We are assuming either a single hazardous substance or a
separate and independent rulemaking process for every substance. We make this simpli-
fying assumption because of our lack of understanding of how the choice of the number
of substances to regulate affects the selection of standards and of category assignments.
In addition, note that not all standards are written in the format we describe. For example,
some pollution standards may be denominated in units of allowable discharge per day,
independent of the firm's production rate.

[40] Our use of the word "firm" refers to each separate source of hazardous substances,
so multiplant enterprises are treated as multiple firms in our formulation. In practice, the
distinction is even more complex because some plants produce multiple products, which
may be regulated under separate categories. Our model formulation assumes that each
plant is regulated as if it consists of many separate plants, each producing a different
product.

To specify the agency decision problem more precisely, we need to define another binary variable $\tau_i$ to count how many categories are actually used, that is, have firms assigned to them. If any firms are assigned to category $i$, then $\tau_i$ equals 1; otherwise $\tau_i$ is 0. Thus, the agency divides an industry into a number of categories

$$k = \sum_{i=1}^{m} \tau_i \tag{4A.3}$$

In addition, by definition the $\tau_i$ satisfy

$$m\tau_i \geq \sum_{j=1}^{m} \delta_{ij} \qquad i = 1, 2, \ldots, m \tag{4A.4}$$

because if one or more firms are assigned to category $i$, then equation (4A.4) requires that $\tau_i$ equal 1, and if no firms are assigned to category $i$, the maximization of $M$ in equation (4A.1) requires that $\tau_i$ equal 0 in order that $k$ be as small as possible. We must also constrain each firm to be assigned to one category, thus

$$\sum_{i=1}^{m} \delta_{ij} = 1 \qquad j = 1, 2, \ldots, m \tag{4A.5}$$

The probability of agency support $f_q(\cdot)$ by a citizen $q$ affected by improvements in the ambient, workplace, or home environment depends on the damage $v_r$ in the region $r$ where the citizen lives or works. We assume damage to be a public bad, so $v_r$ depends on the total level of hazardous substance discharge $s_r$ by all firms in region $r$. Let $\gamma_{rj}$ be a binary variable which equals 1 if firm $j$ discharges in region $r$ and equals 0 otherwise. Then

$$S_r = \sum_{j=1}^{m} \gamma_{rj} \, s^j \tag{4A.6}$$

Similarly, define the binary variable $\alpha_{rq}$ to take on the value 1 if citizen $q$ lives in region $r$ and equal 0 otherwise. That person's probability of support must then equal

$$f_q\left( \sum_{r=1}^{R} \alpha_{rq} \, v_r\,(S_r) \right)$$

when the country is divided into $R$ mutually exclusive, but exhaustive, regions.

We can now write the regulatory agency's decision problem, which is to select category standards $s_i$ and firm assignments $\delta_{ij}$ to maximize

$$
M = \sum_{q=1}^{n} f_q \left( \sum_{r=1}^{R} \alpha_{rq}\, v_r\, (S_r) \right)
$$
$$
- \sum_{j=1}^{m} e_j\, h_j \left( t_j(s^1, \ldots, s^j, \ldots, s^m) \right) - ck \tag{4A.7}
$$

subject to the constraints in equations (4A.2) through (4A.6).

Given the choice of firm assignments $\delta_{ij}$, the category standards $s$ must satisfy the first-order conditions:

$$
0 = \sum_{q=1}^{n} f_q' \sum_{r=1}^{R} \alpha_{rq} v_r' \left( \sum_{j=1}^{m} \gamma_{rj} \delta_{ij} p_j \right)
$$
$$
- \sum_{j=1}^{m} e_j h_j' \left[ \sum_{u=1}^{m} \delta_{iu}(\partial t_j/\partial s^u) p_u \right] \qquad i = 1, 2, \ldots, m \tag{4A.8}
$$

We assume that a higher regional discharge (larger $S_r$) leads to increased damage ($v_r' > 0$), which causes less citizen support ($f_q' < 0$). We also assume that a firm $j$ in a category with a standard that is weakened (higher $s$) is made more profitable because that firm's standard is relaxed ($\partial t_j/\partial s^j < 0$), but its profits are reduced to the extent that other firms in the same category receive weaker standards and thus are made more profitable competitors ($\partial t_j/\partial s^u > 0$ for $u \neq j$). Finally, higher costs to the firm as a consequence of standard changes are assumed to increase the firm's opposition to the agency ($h_j' > 0$).

Since the expression in the parentheses in equation (4A.8) gives the production by firms in region $r$ and category $i$, the first term in equation (4A.8) is the increase in citizen support for an agency due to a unit strengthening (decrease) in category standard $s_i$. The expression in brackets gives the profit reduction, or added cost, for firm $j$ of a unit tightening (decrease) of category standard $s_i$. Thus, the second term in equation (4A.8) is the increase in industry opposition to the agency caused by the tighter category standard $s_i$. Condition (4A.8) then provides the not surprising result that the regulatory agency sets all category standards to equate at the margin the increase in the citizen support with the decrease in support from firms.

The necessary condition that characterizes the assignments $\delta_{ij}$ of firms to categories can be stated, but it is more complex and of limited use. This condition requires that given the subcategory standards $s_i$, any firm $w$ should be assigned to subcategory $y$ (that is, $\delta_{yw} = 1$), if for all subcategories $z \neq y$,

$$
\left[ \sum_q f_q \left( \sum_r \alpha_{rq} v_r(S_r) \right) \Big|_{\delta_{yw} = 1} - \sum_q f_q \left( \sum_r \alpha_{rq} v(S_r) \right) \Big|_{\delta_{zw} = 1} \right]
$$

$$
\geq \left\{ \left[ \sum_j e_j h_j' \left( \sum_u \delta_{iu} (\partial t_j / \partial s^u) p_u \right) \Big|_{\delta_{yw} = 1} \right] \right.
$$

$$
- \left[ \sum_j e_j h_j' \left( \sum_u \delta_{iu} (\partial t_j / \partial s^u) p_u \right) \Big|_{\delta_{zw} = 1} \right]
$$

$$
\left. - \left[ c \left( k \Big|_{\delta_{yw} = 1} - k \Big|_{\delta_{zw} = 1} \right) \right] \right\} \qquad (4A.9)
$$

In words, it requires that for any firm $w$ assigned to category $y$, assigning the firm to any other category $z$ reduces (increases) support from citizens more (less) than it reduces (increases) opposition from firms, adjusted for any difference in administrative cost because of a change in the number of categories used.[41]

---

[41] Note that conditions (4A.8) and (4A.9) provide only necessary conditions for maximizing agency support (4A.7). Given specific functional forms for the $f_q$, $v_r$, $h_j$, and $t_j$ functions and specific values for the parameters $\alpha_{rq}$, $\gamma_{rj}$, $p_j$, and $e_j$, we can suggest a rough algorithm for solving for the optimal values of $\delta_{ij}$, but it is rather simplistic and would only be reasonable to apply to problems with small values of $m$:

1.  Select an interval and identify all possible values of the multiple $(s_1, s_2, \ldots, s_m)$ whose component elements differ by multiples of the interval. Eliminate all multiples with component values that are equal (including equal to 0), since the agency wishing to minimize category costs $ck$ would never assign firms to more than one category with the same standard.

2.  For each multiple, solve the integer-programming problem presented by (4A.7) and constraints (4A.2) through (4A.6).

3.  Select the largest solution value of $M$ and interpolate around the associated multiple with a finer grid of multiple values.

It is interesting to observe that most regulatory agencies do not follow the spirit of this algorithm. They generally take the opposite approach of first assigning firms to categories and then selecting category standards.

*The Categorization Model*

Ideally, we would like to derive hypotheses from the general model above that relate characteristics of firms in an industry and their regulatory environment to the degree to which the industry is partitioned into separate categories, each with separate standards. This approach was taken in the body of the chapter to develop hypotheses about the stringency of regulations from the stringency decision model. Unfortunately, the complexity of the category assignment problem as posed in this general model precludes comparative statics analysis of the factors that affect the degree of categorization.[42] Relating characteristics of firms in an industry group to the degree of industry categorization requires a less complex model of the categorization decision, as well as a meaningful measure of the degree of categorization. This subsection reduces the general model to a less complex form amenable to comparative statics analysis similar to that performed with the earlier stringency decision model. Recognize, however, that a considerable simplification of the general model must be made in order to be able to carry out the comparative statics analysis.

Using the same approach that was taken with the model of the stringency decision, we now derive a table of comparative statics results that relate changes in the parameters of the model to changes in the degree of industry categorization. By associating parameters in the table with measurable characteristics of an industry and its regulatory environment (as was done earlier in this chapter), the table enables us to generate potentially testable hypotheses about the factors that explain the diversity in the degree of categorization across industries that are regulated by the same agency.[43] In contrast to our successful data collection effort to test hypotheses about stringency decisions, we were unable to collect the data necessary to test the hypotheses about the categorization decision. Part of this problem was conceptual and involved the difficulty in deriving a meaningful measure of the degree of industry categorization. This problem is discussed further in appendix 5-B. Further research is needed in order to submit these hypotheses to an empirical test.

Let us return to the model of the stringency decision and the regulatory problem posed in equation (4.3). Now add a second component $k$ to

---

[42] The binary nature of the category allocation decision variable for each firm creates the main impediment to this form of analysis.

[43] The comparative statics results in table 4-1 continue to hold, so all the earlier hypotheses that were generated about factors that affect the stringency of standards still follow from this model.

the compliance cost function, where $k$ is the number of categories into which the $m$ firms in the industry are divided. Thus, $t(s/m,k)$ is the cost to a representative firm in the industry of complying with a total limitation $s$ on hazardous substances with $k$ categories of firms each receiving separate standards. For example, $s$ could represent the total number of accidents allowed per year under a set of OSHA standards for the industry; or $s$ could be the daily amount of sulfur dioxide that EPA allows an industry to discharge into the air. Obviously, $k$ must be less than or equal to $m$.

Again, we assume that industry costs decrease at a decreasing rate with the level of the standard ($t_1 < 0$ and $t_{11} > 0$). Adding more categories allows the total limitation $s$ to be spread among firms in a way that more closely approximates the minimum-cost allocation. Thus, industry costs also decrease at a decreasing rate with the number of categories ($t_2 < 0$ and $t_{22} > 0$). In addition, increasing the number of subcategories decreases the effectiveness in cost reduction of an increase in the industry standard ($t_{21} < 0$).[44]

The regulatory agency has two mechanisms for reducing industry opposition to its actions through decreasing the compliance cost $t(s/m,k)$. First, it can weaken the standards imposed on all firms in the industry by raising $s$, the industry standard. Second, the agency can maintain the same level of the total standard for the industry $s$, but it can create additional categories for those firms with particularly high compliance costs and weaken the standards in those subcategories in order to reduce those firms' strong opposition to the industry standard $s$. Of course, if the total volume of hazard or risk exposure created by the industry is held fixed, then relaxing the standards for some categories requires tightening them for others.

Although adding more categories allows the regulatory agency to reduce industry opposition to its standards, it also imposes administrative costs, $ck$, which for simplicity we will assume to be proportional to the number of categories (where $c > 0$). Making trade-offs between external support and administrative costs requires the agency to translate from the currency of its internal administrative costs to that of external support.

The agency's objective function in equation (4.1) must now be modified to incorporate (1) the effect of the degree of categorization, $k$, on the compliance cost function $t(s/m,k)$, (2) the administrative costs of

---

[44] The $t_{21} > 0$ assumption is consistent with (but not necessary for) the second-order conditions for the agency optimization problem to be analyzed below. Also, if $t$ were to be so regular as to be linear homogeneous, then necessarily $t_{21} > 0$.

creating separate categories, $ck$, and (3) the four parameters $a$, $b$, $r$, and $p$. Thus, the new objective function is

$$M = nbf[pv(s)] - meah[rt(s/m,k)] - ck \qquad (4A.10)$$

A regulatory agency attempting to maximize its net external support $M$ selects the stringency of the optimal industry standard $s^*$ and the optimal number of subcategories $k^*$ to satisfy

$$0 = nbf'[pv(s^*)]pv'(s^*) - eah'[rt(s^*/m,k^*)]rt_1(s^*/m,k^*) \qquad (4A.11)$$

and

$$c = -meah'[rt(s^*/m,k^*)]rt_2(s^*/m,k^*) \qquad (4A.12)$$

Comparative statics analysis of the stringency of standard $s^*$ yields the same results as for the stringency decision model without categorization (see table 4-1). Table 4A-1 summarizes the effects of parameter changes on the degree of categorization $k^*$. Before deriving these results, consider their implications for the degree of industry categorization. An agency reacts to an increase in firms' sensitivity to increases in compliance costs (parameter $a$) by adding categories to mitigate the costs. When the number of people damaged (parameter $n$) or their sensitivity to damage (parameter $b$) are increased, then the agency responds by tightening the standard but cushioning the associated costs on vocal or politically powerful segments of the industry by creating new categories with relatively weaker standards. If the political clout of the industry (parameter $e$) increases relative to that of citizens, then the agency reacts by reducing industry compliance costs both through weaker standards

TABLE 4A-1.   COMPARATIVE STATIC EFFECTS ON
THE DEGREE OF INDUSTRY CATEGORIZATION

| Parameter increased | Sign of associated change in $k$ |
|---|---|
| $n$ (number of citizens)[a] | + |
| $e$ (relative power of firms) | + |
| $a$ (probability of firm opposition) | + |
| $b$ (probability of citizen opposition)[a] | + |
| $r$ (compliance cost) | + |
| $p$ (damage)[a] | + |
| $c$ (categorization cost) | − |

[a] These three parametric effects all take the sign of $(-M_{sk})$, which we assume to be positive.

and narrower categories. Similarly, for an industry with a higher compliance cost curve (that is, a larger value of parameter $r$), the agency
provides weaker standards and more categories. For an industry causing
more damage, the agency grants citizens more stringent standards, but
lightens the burden on some industry segments by adding categories.
Finally, higher subcategory administrative costs (parameter $c$) lead the
agency to reduce these costs by creating fewer categories.

We now derive the results given in table 4A-1. The first-order conditions (4A.11) and (4A.12) yield the following matrix of second-order
partial derivatives:

$$D = \begin{bmatrix} M_{ss} & M_{sk} \\ M_{ks} & M_{kk} \end{bmatrix} \tag{4A.13}$$

Given the sign assumptions in the text,

$$\begin{aligned} M_{ss} &= (nbp)(f'v'' + pv'^2f'') \\ &- (ear)[h't_{11}/m + rt_1^2h''/m] < 0 \end{aligned} \tag{4A.14}$$

and

$$M_{kk} = (-mear)h't_{22} + h''t_2^2r) < 0 \tag{4A.15}$$

Also, $t_{21} = t_{12}$, so that

$$M_{ks} = M_{sk} = (-ear)(h't_{21} + h''rt_1t_2) \tag{4A.16}$$

For the solution to equation (4A.10) to be a maximum requires $M_{ss}$
$< 0$, $M_{kk} < 0$, and $D = M_{ss}M_{kk} - M_{ks}^2 > 0$. From equations (4A.14)
and (4A.15), the first two conditions are satisfied. For the third condition
to be met, we assume that direct second-order effects outweigh cross-
partial effects. If $t$ is linear homogeneous, then it can be shown that the
third condition holds.

Using Cramer's Rule and the total derivatives of the first-order conditions, we derive the signs in tables 4-1 and 4A-1.

$$\frac{ds}{dn} = (bf'pv'mear)(h't_{22} + h''t_2^2r)/|D| < 0$$

$$\frac{ds}{de} = (ea^2h'^2r^2)(t_{21}t_2 - t_1t_{22})/|D| > 0$$

$$\frac{ds}{da} = (e^2ah'^2r^2)(t_{21}t_2 - t_1t_{22})/|D| > 0$$

$$\frac{ds}{db} = (nf'pv'mear)(h't_{22} + h''t_2{}^2r)/|D| < 0$$

$$\frac{ds}{dr} = (-e^2a^2rh')(h' + rh''t)\,(t_1t_{22} - t_2t_{22})/|D| > 0$$

$$\frac{ds}{dp} = (nbv'mear)(f' + pvf'')\,(h't_{22} + h''t_2{}^2r)/|D| < 0$$

$$\frac{ds}{dc} = (-M_{sk})/|D| > 0, \text{ if } M_{sk} \text{ assumed negative}$$

$$\frac{dk}{dn} = (-M_{sk})(-nf'pv')/|D| > 0, \text{ if } M_{sk} \text{ assumed negative}$$

$$\frac{dk}{de} = (ea^2r^2h'^2)(t_{21}t_1 - t_{11}t_2)/|D| > 0$$

$$\frac{dk}{da} = (nbpeh'rt_2)(f'v'' + pv'^2f'') + (e^2ar^2h'^2)(t_{21}t_1 - t_{11}t_2)/|D| > 0$$

$$\frac{dk}{db} = (-M_{sk})(-nf'pv')/|D| > 0, \text{ if } M_{sk} \text{ assumed negative}$$

$$\frac{dk}{dr} = \left[(mea)(h' + rh''t)\right]$$
$$\times \left[(t_2nbp)(f'v'' + pv'^2f'') + (earh')(t_{21}t_1 - t_{11}t_2)\right] > 0$$

$$\frac{dk}{dp} = (-M_{sk})(-nbv')(f' + pvf'')/|D| > 0,$$
$$\text{if } M_{sk} \text{ assumed negative}$$

$$\frac{dk}{dc} = M_{ss}/|D| < 0$$

# REFERENCES

Alchian, A., and H. Demsetz. 1972. "Production, Information Costs, and Economic Organization," *American Economic Review* vol. 62, no. 5.

Arnold, R. Douglas. 1979. *Congress and the Bureaucracy: A Theory of Influence* (New Haven, Conn., Yale University Press).

Baron, David P. 1984. "Regulation of Prices and Pollution under Incomplete Information." Working paper (Palo Alto, Calif., Stanford Business School, July).

Beales, Howard, Richard Craswell, and Steven C. Salop. 1981. "The Efficient Regulation of Consumer Information," *Journal of Law and Economics* vol. 24 (December), pp. 491–545.

Committee on Government Affairs, U.S. Senate, 1977. *Study on Federal Regulation* vol. V (Washington, D.C., U.S. Government Printing Office).

Cyert, Richard M., and James G. March. 1963. *Behavioral Theory of the Firm* (Englewood Cliffs, N.J., Prentice-Hall).

Dahl, Robert, and Charles Lindblom. 1953. *Politics, Economics and Welfare* (New York, Harper).

Downs, Anthony. 1967. *Inside Bureaucracy* (Boston, Little, Brown).

Fiorina, Morris P. 1981. "Legislative Choice of Regulatory Forms: Legal Process or Administrative Process?" Social science working paper 387 (Pasadena, Calif., California Institute of Technology, May).

———, and Roger G. Noll. 1978. "Voters, Bureaucrats and Legislators: A Rational Choice Perspective on the Growth of Bureacracy," *Journal of Public Economics* vol. 9 (April).

Groves, Theodore, and Martin Loeb. 1976. "Incentives and Public Inputs," *Journal of Public Economics* vol. 4, no. 3, pp. 211–226.

Henderson, J. M., and R. E. Quandt. 1971. *Microeconomic Theory* (New York, McGraw-Hill).

Holden, Matthew, Jr. 1966. *Pollution Control as a Bargaining Process: An Essay on Regulatory Decision-Making* (Ithaca, N.Y., Cornell University Water Resources Center).

Ingram, Helen M., and Scott J. Ullery. 1977. "Public Participation in Environmental Decision-Making: Substance or Illusion?" in W. K. Derrick Sewell and J. T. Coppock, eds., *Public Participation in Planning* (New York, John Wiley & Sons).

Kolko, Gabriel. 1965. *Railroads and Regulation: 1877–1916* (Princeton, N.J., Princeton University Press).

Levine, Michael E. 1981. "Revisionism Revised? Airline Deregulation and the Public Interest," *Law and Contemporary Problems* vol. 44, no. 1, pp. 179–195.

Loeb, Martin, and Wesley A. Magat. 1979. "A Decentralized Method of Public Utility Regulation," *Journal of Law and Economics* vol. 22 (October).

March, James G., and Herbert A. Simon. 1958. *Organizations* (New York, John Wiley & Sons).

Marschak, J., and R. Radnor. 1972. *Economic Theory of Teams* (New Haven, Conn., Yale University Press).

Mirrlees, J. A. 1976. "The Optimal Structure of Incentives and Authority Within an Organization," *The Bell Journal of Economics* vol. 7, no. 1.

Mitnick, Barry M. 1980. *The Political Economy of Regulation: Creating, Designing and Removing Regulatory Forms* (New York, Columbia University Press).

Moe, Terry M. 1982. "Regulatory Performance and Presidential Administration," *American Journal of Political Science* vol. 26, no. 2.

Nagal, Jack H. 1975. *The Descriptive Analysis of Power* (New Haven, Conn., Yale University Press).

Noll, Roger G. 1971. *Reforming Regulation: Studies in the Regulation of Economic Activity* (Washington, D.C., The Brookings Institution).

———. 1976. "Government Administrative Behavior and Private Sector Response: A Multidisciplinary Survey," Social science working paper 62 (Pasadena, Calif., California Institute of Technology).

Peltzman, Sam. 1976. "Toward a More General Theory of Regulation," *Journal of Law and Economics* vol. 19, no. 2, pp. 211–240.

Porter, Michael E., and Jeffrey F. Sagansky. 1976. "Information Politics and Economic Analysis: The Regulatory Decision Process in the Air Freight Cases," *Public Policy* vol. 24, no. 2 (Spring), pp. 263–307.

President's Advisory Council on Executive Organization. 1971. *A New Regulatory Framework: Report on Selected Independent Regulatory Agencies* (Washington, D.C., U.S. Government Printing Office).

Roberts, Marc J., and Jeremy S. Bluhm. 1981. *The Choice of Power: Utilities Face the Environmental Challenge* (Cambridge, Mass., Harvard University Press).

Sappington, David. 1983. "Optimal Regulation of a Multiproduct Monopoly with Unknown Technological Capabilities," *Bell Journal of Economics* vol. 14, no. 2, pp. 453–463.

Simon, H. A. 1957. *Models of Man* (New York, John Wiley and Sons).

Sonstelie, John C., and Paul R. Portney. 1983. "Truth or Consequences: Cost Revelation and Regulation," *Journal of Policy Analysis and Management* vol. 2, no. 2, pp. 280–284.

Spence, A. M. 1975. "The Economics of Internal Organization: An Introduction," Papers from a symposium on the Economics of Internal Organization, *Bell Journal of Economics* vol. 6, no. 1.

Stewart, Richard B. 1975. "The Reformation of American Administrative Law," *Harvard Law Review* vol. 88, no. 8, pp. 1669–1813.

Stigler, George J. 1971. "The Theory of Economic Regulation," *Bell Journal of Economics* vol. 2, no. 1, pp. 3–21.

Stiglitz, Joseph E. 1979. "Equilibrium in Product Markets with Imperfect Information," *American Economic Review* vol. 69, no. 2, pp. 339–345.

Vogelsang, Ingo, and Jorg Finsinger. 1979. "A Regulatory Adjustment Process for Optimal Pricing by Multiproduct Monopoly Firms," *Bell Journal of Economics* vol. 10, no. 1, pp. 157–171.

Weingast, Barry R. 1984. "The Congressional-Bureaucratic System: A Principal Agent Perspective with Applications to the SEC," *Public Choice* vol. 44, no. 1, pp. 147–191.

———, and Mark J. Moran. 1983. "Bureaucratic Discretion or Congressional Control: Regulatory Policymaking by the Federal Trade Commission," *Journal of Political Economy* vol. 91, no. 5, pp. 765–800.

Williamson, Oliver E. 1964. *The Economics of Discretionary Behavior: Man-*

*agerial Objectives in a Theory of the Firm* (Englewood Cliffs, N.J., Prentice-Hall).
———. 1976. "Franchise Bidding for Natural Monopolies—in General with Respect to CATV," *Bell Journal of Economics* vol. 7, no. 1, pp. 3–104.
Wilson, James Q. 1974. "The Politics of Regulation," in James W. McKie, ed., *Social Responsibility and the Business Predicament* (Washington, D.C., The Brookings Institution).

# 5
# Methodological Issues

Applying the revealed preference approach to understanding the behavior of a regulatory agency is far from straightforward. Any application of statistical approaches to rulemaking must resolve a variety of methodological issues. The processes for setting effluent guidelines, for example, have two important characteristics. They consist of several closely related stages and they yield continuous decisions, such as the levels of standards, rather than discrete decisions, such as whether or not to regulate a substance at all.

## STRUCTURE OF THE MODEL

The Best Practicable Technology (BPT) rulemaking process can be separated into three key stages: the issuance of contractor recommendations (denoted by $C$), the publication of proposed rules (denoted by $P$), and the promulgation of final rules (denoted by $R$). Contractors issued a report containing their recommended standards along with one or more abatement technologies to meet them. After EPA received comments and an economic analysis report on the aggregate effects of the contractor's recommended rules, the Effluent Guidelines Division issued proposed rules, which were published in the *Federal Register*. Then, after repeating the comment and economic analysis process on the proposed rules, EPA issued its promulgated rules.

These three stages were not independent, for the stringency of the contractor standards affected subsequent factors, such as the number of industry comments and the employment effects forecast in the economic analysis reports, which in turn affected the proposed standards. The stringency of the proposed standards affected the promulgated standards. Thus, for purposes of econometric analysis, this rulemaking process must be modeled as an interdependent system.

Of all types of interdependent systems, the recursive system—the simplest in terms of statistical estimation—appears to fit the BPT process best. Such a system requires that the endogenous (dependent) variables in an equation describing one stage be exogenous (independent) variables in equations describing subsequent stages. This condition is met simply because of the temporal nature of the BPT process; that is, the different sets of standards cannot possibly be determined simultaneously, although the earlier standards affect the later standards.

From the equations for a general linear recursive system, it can be shown that the BPT rulemaking process for each pollutant is characterized by the following three equations:

$$S^C = a_o + a_1x_1 + a_2x_2 + \ldots + a_Mx_M + \varepsilon_1 \tag{5.1}$$

$$\Delta S^{C,P} = b_o + b_1x_1 + b_2x_2 + \ldots + b_Mx_M$$
$$+ b_{M+1} S^C + \varepsilon_2 \tag{5.2}$$

$$\Delta S^{P,R} = c_o + c_1x_1 + c_2x_2 + \ldots + c_Mx_M$$
$$+ c_{M+1} S^C + c_{M+2} \Delta S^{C,P} + \varepsilon_3 \tag{5.3}$$

The variables $S^C$, $S^P$, and $S^R$ are the discharge standards for the pollutant issued during the contractor $(C)$, proposed $(P)$, and promulgated $(R)$ stages of the rulemaking process. The revision in the standard from the contractor to the proposed stages is denoted by $\Delta S^{C,P}$, and the change in the standard from the proposed to the promulgated stage is denoted by $\Delta S^{P,R}$. The explanatory variables $x_1, x_2, \ldots, x_M$ may differ in each equation, and some $x_i$ may be relevant to only certain pollutants.

The errors $\varepsilon_1$, $\varepsilon_2$, and $\varepsilon_3$ are assumed to be normally distributed with mean zero. In addition, the use of ordinary least squares to estimate the parameters of this recursive system and obtain efficient estimates requires the assumption that no correlation of disturbances is present across equations. Fortunately, although errors in measurement involved in explaining variation in the contractor standards (equation 5.1) may be related to those arising from a regression equation explaining variation in the proposed standards, they are unlikely to be related to those involved in explaining variation in changes in the standards (equations

5.2 and 5.3). Other sources of cross-equation correlation, such as omitted variables, might still be present, however. Use of a Hausman test (Hausman, 1978) confirmed the validity of the assumptions of zero correlation of disturbances and of orthogonality.

Without changing the spirit of the model, the dependent variables in equations (5.2) and (5.3) can be expressed in terms of percentage changes. This modification is advantageous because a percentage change variable is unit free, thus eliminating the problem of comparing standards written in noncomparable units for these two equations.

Although we have argued that the basic structure of the BPT process is recursive, some individual elements are not. In particular, the predictions of economic impacts contained in the economic analysis reports may have been determined simultaneously with the proposed standards. These reports were supposed to provide the economic implications of complying with the contractor's recommended standards. Therefore, they would have helped to determine changes to be made in these recommendations before they were issued as proposed standards. The economic analysis reports were often updated before publication, however, to reflect impacts associated with the proposed rules. In contrast, the second economic analysis report was issued with the promulgated rules and was, from the start, to reflect economic impacts of those final rules.

Given these facts about the two economic analysis reports and the difficulties of modeling such a simultaneous system, the economic projections issued with the proposed standards are treated as independent variables to explain changes in standards from the proposal to promulgation stages; the first economic analysis report was not used to generate explanatory variables for the changes in standards from the contractor to the proposal stages.

After selecting the model's structure, the question was which pollutants to model. This decision was complicated, because the treatment technologies often remove more than one pollutant. Thus, the structural equations explaining the level of standards for different pollutants in the same industry were not likely to be independent.

Under certain conditions, these interdependencies would be easy to disentangle. If abatement technologies were designed in such a way that the standard for one pollutant was always binding—in the sense that a relaxation or tightening of this standard alone would change compliance costs (or damages) and variations in the standards for any other pollutant would not affect compliance costs or damages—then the equation system could focus on this one pollutant. Alternatively, if all treatment technologies removed all pollutants in identical and fixed proportions, a set of standards on any pollutant could be chosen as the focus of the equation system. Unfortunately, treatment technologies rarely exhibit either of these two convenient characteristics.

To deal with this problem, we estimated the reduced-form equations for the two pollutants that dominated the rulemaking process—biological oxygen demand (BOD) and total suspended solids (TSS). Standards on these pollutants vastly outnumber those on the others, and interest in controlling BOD and TSS appears to have been paramount in the legislative history.[1] Use of the reduced-form equations yields unbiased coefficient estimates. Because no well-developed theory exists for explaining the structural relationship between the BOD standards for an industry category and its TSS standards, attempting to estimate the underlying structural equations seemed futile. For purposes of classifying results, if a variable is significant in a BOD regression but not in a TSS regression, or vice versa, that variable is treated as having a significant effect on the BPT rulemaking process.

## SCOPE OF THE STUDY

The process of controlling water pollution from industrial sources under the Federal Water Pollution Control Act Amendments of 1972 was doubtlessly influenced by many factors, from legislative activities predating the act to monitoring and enforcement activities carried out at the state and federal levels. In between, EPA promulgated Best Practicable Technology (BPT) standards, courts were asked to review the promulgated rules, and permits based on the standards were written for specific facilities. These stages in the life of the regulations had an effect on the process.

The rulemaking, or standard-setting, process—EPA's issuance of the contractor's recommended standards, the proposed standards, and the promulgated rules—yielded such rich variation in the stringency of the standards that it provides an ideal data set for statistical analysis. In contrast, the legislative process concerning BPT standards yielded few outcomes, and these results, such as the choice of the technical approach to rulemaking, are difficult to quantify. The outcomes of permit and

---

[1] In many cases, one or the other of these pollutants is clearly binding, that is, the standards on other pollutants are written with reference to the technology chosen to deliver a desired amount of BOD or TSS removal. Thus, if we can explain the variation in these two pollutants taken together, we will have explained most of the variation in the stringency of all pollutants across subcategories for the entire BPT process.

enforcement activities—permitted and actual emissions—do not pose this quantification problem, provided the information is available. The decentralized nature of permit and enforcement activities and still un-resolved definitional questions have precluded establishing a national data base of permitted and actual emissions. The judicial actions and opinions applicable to each set of industry rules probably could be char-acterized for statistical analysis, but such a data collection effort and application must await future research.

The focus on standard setting to the exclusion of other stages in the rulemaking process does place some limitations on the types of conclu-sions that can be drawn. The determinants of the final outcomes—the set of permits to which plants were subject and the actual emissions resulting from the permitting and enforcement efforts—cannot be eval-uated. An analysis of such outcomes would be necessary to reach con-clusions about the social effects of the BPT process. This focus also bypasses another potentially important nuance of the process—the ef-fect that forward-looking (or strategic) behavior by bureaucrats has on the outcomes of the rulemaking process. For instance, a court remand of particular rules may influence future rules by changing the procedures and information requirements that bureaucrats feel they need to win court tests. Without including the court outcomes in the rulemaking process model, the extent of such effects cannot be examined.

This study also fixes a temporal boundary around the BPT process. With standards set on Group One and Group Two industries over a five-year period (1973–1977), underlying economic and political con-ditions, as well as institutional learning and key, top-level personnel changes, could have become major factors in explaining why rules were set as they were. But such conditions are numerous and difficult to quantify and might well distort findings of a study encompassing such a long period of time. Restricting the study to rules set for Group One industries shortens the study period so that it ends before the change in administration in 1976. In addition, and more important, using a limited set of industries reduces the extensive data development problem associated with creating an integrated, consistent data base.

## DATA DEVELOPMENT

Developing the data bases for statistical description and hypothesis tests of the BPT process involved many decisions concerning the broad issues of sample size and level of aggregation, as well as the specific issues of how to define and measure the dependent and independent variables.

Resolving these issues raised other generic types of data problems such as where to find data, how to resolve inconsistencies in data from different sources, how to handle data omissions, and how to solve the problems of measurement units and aggregation.

## The Data Base: Sampling and Aggregation

Building on the concept of an industry subcategory, a data base of BPT standards was created for all subcategories within each of the forty-nine Group One industries. The data base includes standards for every subcategory at each stage of the process, as well as codes that enable the standards to be tracked across stages as subcategories are combined, split, dropped, and created. The data file covered standards on ninety-six pollutants in both thirty-day average and one-day maximum form. The reason for subcategorization given in the relevant EPA documentation is also included.

This data base, without modification or addition, allows the BPT process to be described statistically. It permits rather crude, cross-subcategory and cross-industry comparisons of standard stringency, changes in stringency, and changes in the number of subcategories. In some instances, it even permits some hypothesis testing (for example, as to whether the standards were based upon cost-effectiveness). Results of such comparisons are described in chapter 6.

In order to perform more complicated hypothesis tests about the determinants of the stringency of the standards, a sample of twenty-three Group One industries was chosen, within which standards were written for 268 subcategories (table 5-1). Because data on some of the suspected determinants were unavailable at the level of EPA-defined subcategories, some of the sample subcategories were aggregated, leaving a total of 106 subcategories.[2] For these subcategories, data were collected describing industry and rulemaking characteristics that were hypothesized to influence the BPT standard-setting process.

The sample of twenty-three Group One industries was not chosen randomly. First, scarcity of time and funding required eliminating industries with extremely involved standards—those that would be difficult and costly to express in a way amenable to econometric testing. Second, because EPA's subcategorization decision could not be ex-

---

[2] Nine EPA-defined subcategories were dropped due to data limitations or inappropriate subcategory definitions that made cross-comparisons impossible (for example, a separate pollutant limitation for rainfall and runoff effects in the cement manufacturing industry could not be expressed in specific mg/l or percentage removal terms). Appendix 5-A provides a detailed description of the data base.

TABLE 5-1.  GROUP ONE INDUSTRIES

| Industries included in the sample | Other industries |
|---|---|
| Builders paper and board mills | Asbestos manufacturing I |
| Cane sugar refining (Sugar I) | Asbestos manufacturing II |
| Cane sugar processing (Sugar II) | Dairy product processing |
| Cement manufacturing | Electroplating I |
| Canned and preserved fruits and | Electroplating II |
| vegetables processing I | Feedlots |
| Canned and preserved fruits and | Ferroalloys I |
| vegetables processing II | Ferroalloys II |
| Flat glass manufacturing (Glass I) | Basic fertilizer chemicals |
| Pressed and blown glass | (Fertilizer I) |
| manufacturing (Glass II) | Formulated fertilizer (Fertilizer II) |
| Major inorganic chemicals | Grain mills I |
| (Inorganics I) | Grain mills II |
| Leather tanning and finishing | Inorganic chemicals II |
| Meat packing (Meats I) | Iron and steel manufacturing I |
| Meat processing (Meats II) | Iron and steel manufacturing II |
| Poultry processing (Meats III) | Nonferrous metals I |
| Major organic chemicals (Organics I) | Nonferrous metals II |
| Organic chemicals II | Petroleum refining |
| Pulp, paper and paperboard I | Plastics and synthetics I |
| Phosphorus derived chemicals | Plastics and synthetics II |
| (Phosphates I) | Pulp, paper and paperboard II |
| Other nonfertilizer phosphate | Canned and preserved seafood |
| chemicals (Phosphates II) | processing I |
| Tire and synthetic rubber (Rubber I) | Canned and preserved seafood |
| Fabricated and reclaimed rubber | processing II |
| (Rubber II) | Soap and detergent manufacturing |
| Textile mills | Steam electric power generating |
| Timber products I | Wood furniture manufacturing |
| Timber products II | |

*Note*: Group One industries were selected from the 28 general industries mandated in the FWPCAA of 1972. Within these industries, EPA divided complicated industries into phases. Regulations for Phase I segments were addressed before those for Phase II.

amined statistically, the few industries with "odd" subcategorization schemes were also eliminated. These included petroleum refining, for which each plant was given a unique subcategory with standards determined by its size and process mix, and soaps and detergents, for which more than sixty categories were created to cover the operations of one plant.[3]

Finally, in order to avoid the costs of collecting data on intervenor characteristics, the sample was limited to industries without known or

[3] We also tried to limit our sample to industry sectors with error-free rulemaking. Changes made to the standards because of EPA's errors are uninteresting analytically, unless the reasons for errors are systematic and related to variables that can be controlled or influenced.

reported political influence on their regulatory process. In addition, if the stringency of rulemaking for "apolitical" sectors can be related to other influences, then the explanatory power of the approach used here is likely to be even higher when applied to industry standard-setting processes that generate strong political interests.[4] Interviews with EPA personnel responsible for the rules for each industry in the chosen sample revealed that they felt no direct political pressure from their superiors, the Office of Management and Budget (OMB), Congress, or the White House (although fear of such pressure could have been present). The sample and data base do include industries with politically influential trade associations, however.

Our sample includes industries with rules that were written at different times during the process. Chronologically, the sample begins with glass I, for which standards were first promulgated in February 1974 and ends with the issuance of final pulp and paper II standards in January 1976. In addition, industries with both complex and simple waste-water problems are represented in order to capture any differences in rulemaking based on the relative ease of obtaining and processing technical information. For example, the sample includes inorganic chemicals (Phase I) standards for each of twenty-two separate chemicals and builders paper with only one set of standards.

The aggregation of the 268 subcategories to 106 is unlikely to bias results or lessen their credibility. These aggregations were among those EPA used to present its findings on plant closings, unemployment, and costs attributable to compliance. Although EPA often wrote standards for narrower subcategories, lack of data usually forced the economic analysis to cover several subcategories together. Therefore, the level of aggregation used here should adequately represent the level of detail at which debate and decision making occurred.

Why not aggregate subcategories all the way up to the industry level? Interviews with Effluent Guidelines Division project officers indicated that once the subcategorization decision was made, standards were usually set for each major subcategory independently of other major subcategories in the same industry. Within a number of industries, many subcategories have similar standards for percentage removal or concentration (in milligrams per liter), but this may have been a consequence of the use of similar process technologies or the generation of similar raw waste loads rather than the imposition of a common percentage

---

[4] Of course, the decision rules for industries with heavy political intervention may differ from the revealed decision rules we estimated for our sample. We could test this conjecture without collecting data on the enlarged sample.

removal or concentration standard applied to each subcategory. Sub-categories for which standards were not set independently of one another could have been combined into an aggregate subcategory, but to assume this lack of independence existed would have been to prejudge the decision rules EPA used and would have lost information when sub-category standards were actually set independently.

### Specifying the Dependent Variables

The BOD and TSS standards are linked to the concept of stringency and can be transformed into different units of measurement to facilitate the cross-industry comparisons necessary for applying a statistical ap-proach. Any measure of the stringency of the effluent standards has two desirable characteristics. First, the unit of measurement of the standards must be comparable across all the subcategories for which standards were written. EPA's usual measure, pounds of pollutant per pound of output produced, does not meet this criterion because the products of firms in different subcategories are different (for example, sugar and chickens). Second, the unit of measurement of the standards must be directly related to the costs and benefits of pollution abatement.

Participants in the rulemaking process do not react directly to the standards but to what they imply for costs, if the participant is being controlled, or for damages, if the participant is a beneficiary of regu-lations. The measures of stringency used here focus primarily on costs because those parties interested in the damages (that is, environmental groups) did not participate actively in individual industry rulemaking processes. If the outcomes of the process are defined as the additional costs expected to be incurred by plants in order to comply with the standards issued at each of the three stages, more stringent standards are those implying higher total abatement costs (after normalizing the costs by some measure of subcategory size and assuming that product prices remain constant).[5]

Unfortunately, accurate cost projections were not available. Thus, the total costs of control were not useful as direct measure of the strin-gency of standards. Industry often commented to EPA that the agency's compliance cost estimates were too high, but they rarely indicated how high they were. Occasionally, industry groups financed a study on the economic effects of the standard, but such studies covered few subcat-

---

[5] If prices can be increased, then at least some compliance costs can be passed on to consumers. Changes in profit rates then become an indicator of stringency. Our research indicates that prices were rarely expected to change, because only a small percentage of firms in the industry would be affected by the regulations. Reliable projections of profit rate changes were not available.

egories and their objectivity is open to question. EPA's Economic Analysis Division performed economic impact studies, but the methods used to develop these estimates are suspect and many differences developed between the Effluent Guidelines Division and the Economic Analysis Division about basic assumptions and characteristics of an industry.

Even if the quality of these cost estimates could somehow be improved, the estimates would not necessarily be useful in measuring stringency. First, interviews indicated that EPA's concept of stringency was multidimensional. The cost information was accompanied by projections of plant closings and unemployment that Effluent Guidelines Division staff (most were engineers) and Economic Analysis Division staff (most were MBAs) deemed important indicators of stringency. Thus, closings alone or in some combination with the other economic projections might have been the stringency measure uppermost in the minds of EPA decision makers. Second, these economic projections were provided once, after the contractor's report was issued but before the proposed rules were issued, and again, in updated form, after the proposed rules were issued but before publication of the promulgated rules. Given this timing, such costs might represent the costs associated with the standards in the earlier document or they might correspond to the later of the two documents. Thus, the cost estimates in the economic analysis reports could not be used to obtain separate measures of the stringencies of the standards at each of the three stages in the rulemaking process.

Reliable data on total abatement costs, a direct measure of stringency, were not always available, so it was necessary to find other measures that meet the two desirable characteristics of being comparable across subcategories and being directly related to the total costs and benefits of abatement. Three possible measures are modified forms of the standards published in the contractor reports and in the *Federal Register*. The first two measures are the percentage removal of BOD implied by the BOD standard and the concentration of TSS in wastewater allowed by the TSS standard.[6] These two measures seemed to be reasonable total cost proxies, given BOD and TSS removal technologies, and are stand-

---

[6] A percentage removal standard refers to the percentage of a pollutant that is required to be removed from the raw waste load. A concentration standard refers to the maximum amount of a pollutant (usually expressed in milligrams) allowed to be discharged into receiving waters per liter of waste-water flow. The principal reason for EPA's decision on the units of the standard, according to officials, was the drawbacks of using a percentage removal or concentration standard. EPA feared that stating regulations in concentration terms would encourage dilution while discouraging water conservation. Percentage removal rules would have contained little or no incentive for plants with poor waste management practices to make in-plant changes to reduce influent pollution loading, because conventional treatment technologies, particularly for BOD, give constant percentage removal for a wide range of influent loads.

ard measures of BOD and TSS discharges in engineering practice and in the literature.[7] Standards that require a higher percentage removal of BOD or a lower concentration of TSS in the effluent also yield benefits through improved water quality in most cases.[8]

Because constructing these two measures was difficult and involved the resolution of numerous inconsistencies in the data sources (see appendix 5-B), another stringency variable was developed. This measure is modeled more closely after the published standards and required information that was easier to obtain and less likely to be in error. This additional variable measures the stringency of a standard in pounds of pollutant per dollar of sales (*BSALES* for BOD, *TSALES* for TSS). Because the published standards were nearly always expressed in pounds of pollutant per unit of production, the average price for the product covered by the standard was the only additional piece of data required to transform the standards issued in the *Federal Register* into pounds of pollutant per dollar of sales.

The percentage removal, concentration, and discharge-per-dollar-of-sales transformations of the published standards are used to measure regulatory outcomes at the contractor stage. Changes in the standards are expressed simply by the percentage change in the published standards (usually expressed in pounds of pollutant per unit of production) across any two stages (C to P, P to R, or C to R).[9] This measure should be monotonically related to changes in stringency.

Table 5-2 lists all of the dependent variables, and table 5-3 provides summary statistics about these variables.

*Explanatory Variables*

Chapter 4 identified factors that might have influenced the EPA decisions in the BPT regulatory process. Based on the external signals model, we collected data on three types of explanatory variables: (1) direct signals from participants that were transmitted to the agency; (2) industry characteristics that influenced EPA's perception of the likely responses, or signals, it expected to receive upon issuing standards; and (3) the expected economic impacts of the standards. A fourth class of variables, internal ones, was also identified.

---

[7] Correspondence with Dr. James Patterson, wastewater engineer, Illinois Institute of Technology.

[8] In some unusual cases, such as the discharge of sugar cane wastes into the Pacific Ocean from the coast of the island of Hawaii, the assimilative capacity of the receiving water is so great as to eliminate all dangers from pollution discharge. Thus reductions in effluent discharges yield no benefits.

[9] In calculating the percentage changes, the average of these standards for the stages over which the change occurred is used as the denominator.

TABLE 5-2. DEPENDENT VARIABLES

| | |
|---|---|
| *BODC* | Contractor's recommended BOD standards measured in units of percentage removal from the raw waste load.[a] |
| *BODP* | Proposed BPT standards measured in percentage removal terms. |
| *BODR* | Promulgated BOD standards in percentage removal terms. |
| *BSALESC* | Contractor's recommended BOD standards measured in kg of BOD per dollar of sales (kg BOD/$) in direct-discharging plants. |
| *BSALESR* | Promulgated BOD standards measured in kg BOD/$ in direct-discharging plants. |
| *TSSC* | Contractor's recommended TSS standards measured in milligrams of discharge per liter of wastewater (mgTSS/l). |
| *TSSP* | Proposed TSS standards measured in mgTSS/l). |
| *TSSR* | Promulgated TSS standards measured in mgTSS/l). |
| *TSALESC* | Contractor's recommended standards measured in kgTSS/$. |
| *TSALESR* | Promulgated TSS standards measured in kgTSS/$. |
| *BCP* | The percentage change in BOD standards from stages C to P. |
| *BPR* | The percentage change in BOD standards from stages P to R. |
| *BCR* | The percentage change in BOD standards from stages C to R. |
| *TCP* | The percentage change in TSS standards from stages C to P. |
| *TPR* | The percentage change in TSS standards from stages P to R. |
| *TCR* | The percentage change in TSS standards from stages C to R. |

[a] A transformation, calculated as $LBODC = LOG (BODC/(1 - BODC))$, is necessary to restrict the predicted values of BODC to a range between 0 and 1. Similar logistic transformations were made for *BODP* and *BODR*.

TABLE 5-3. DESCRIPTIVE STATISTICS OF DEPENDENT VARIABLES

| Variable | $N^a$ | Mean | Standard deviation | Minimum | Maximum |
|---|---|---|---|---|---|
| *BODC* | 59 | 88.34 | 19.67 | 0.99 | 100.00 |
| *BODP* | 54 | 91.11 | 14.24 | 0.99 | 100.00 |
| *BODR* | 54 | 85.28 | 21.96 | 0.003 | 100.00 |
| *BSALESC* | 59 | 14.78 | 48.10 | 0 | 344.32 |
| *BSALESR* | 54 | 18.07 | 69.45 | 0 | 508.66 |
| *TSSC* | 102 | 52.29 | 151.84 | 0 | 1428.20 |
| *TSSP* | 97 | 43.43 | 68.64 | 0 | 508.03 |
| *TSSR* | 97 | 78.85 | 136.36 | 0 | 897.83 |
| *TSALESC* | 97 | 12.75 | 65.04 | 0 | 629.95 |
| *TSALESR* | 92 | 18.37 | 59.61 | 0 | 425.35 |
| *BCP* | 54 | 8.30 | 39.65 | − 164.01 | 100.00 |
| *BPR* | 49 | 27.00 | 28.35 | − 27.03 | 95.20 |
| *BCR* | 54 | 24.29 | 61.70 | − 170.12 | 129.41 |
| *TCP* | 97 | 16.39 | 48.47 | − 169.70 | 133.30 |
| *TPR* | 92 | 23.65 | 35.23 | − 51.85 | 126.15 |
| *TCR* | 97 | 33.70 | 65.37 | − 152.90 | 165.85 |

*Note:* These statistics should not be used to make inferences about how the process operated, because they are not weight-averaged. In addition, in the aggregate, they may appear to be inconsistent. For example, from rows 1 and 2, column 3, the average percentage removal of BOD appears to have risen (that is, the standards appear to have been tightened) from stage C (88.34) to stage P (91.11). Yet, the *BCP* mean change shows that standards were weakened by an average of 8.3%. This result is caused by the aggregation procedure. Nevertheless, no individual observations exhibit these inconsistencies (for example, when a subcategory standard exhibits a tightening in percentage removal terms, the percentage change variable is always negative).

[a] $N$ = number of observations.

*Direct External Signals.*   Only three sources of data were available for measuring the strength, frequency, and origin of external signals. The first source is the comments from interested parties (mostly firms and trade associations) that were solicited after release of the contractor report and again after issuance of the proposed rules. The *Federal Register* notices of proposed and promulgated rules provide a second source of data, because EPA published the official agency response to key issues raised in these comments; these notices did not identify the originator of the issue or the frequency with which the issue was raised, but the documents provide a list of commenters. Finally, interviews and secondary sources of information yielded data on external signals transmitted in the BPT process. Project officers and other key EPA personnel provided information about the origin, frequency, and strength of the external signals they received. In addition, representatives of trade associations participating in the process were queried about the size (in terms of budget and number of employees) of their organizations.

The only direct measures of external signals used in this study were the number of comments and the trade association budget. Comments were measured by the number of significant issues raised in the *Federal Register* that argued for weaker standards or for new subcategories with weaker standards (noted as *NC*).[10] This count does not exactly measure either the frequency of the signal or its strength, but it does indicate the degree of pressure brought to bear on the agency to relax particular standards.[11] In particular, signal strength—defined by the logic or persuasiveness of the message—cannot be assessed from the distillation of comments appearing in the *Federal Register*. For this task, the original comments must be referenced and some technique devised to measure persuasiveness. Unfortunately, EPA's collection of comments contains too many gaps to justify such an effort. The frequency of signals also cannot be assessed, because the list of commenters in the *Federal Register* fails to include those engaging in ex parte contact through telephone conversations and meetings. Neither can the number of independent comments be counted, because several different commenters could have made the same comment.

The participating trade association's budgets (noted as *TAB*), which were highly correlated with their number of employees, was used as a

---

[10] Issues that argued for tighter standards were rarely raised.

[11] By covering two periods in the rulemaking process (C-P and P-R) and separating the comments concerning each of the two pollutants (BOD and TSS), we defined six separate comment variables concerned with issues of a more general nature. These were added to the counts for each pollutant because commenters often did not single out a specific pollutant in their letters. (See table 5-4.)

crude measure of their signal strength. If a trade association had an office in Washington, D. C., it might have superior access to information or people. This variable could not be used, however, because every industry in the sample had at least one trade association located in Washington at the time of the BPT standard-setting process.

*Industry Characteristics.*   Guided by the external signals theory and the availability of data, seven industry and subcategory characteristics were used in the analysis—the four-firm concentration ratio ($CR4$), the rate of return on equity ($ROR$), the number of small plants ($NSP$), the number and percentage of plants at risk ($NPAR$ and $PPAR$), the raw waste load ($RAW$), and the wastewater flow ($FLOW$).

The single most difficult problem with obtaining data for these variables was determining whether the available data applied to the appropriate aggregation of plants. The degree of aggregation used to present these variables, whether they were found in economic analysis reports, census publications, or other sources, often did not coincide with the subcategorization scheme used by the Effluent Guidelines Division to group similar firms and plants. Furthermore, EPA was under extreme time pressure and made little attempt to adjust the available data to fit the level of aggregation used for setting standards. For example, the contractor might have found the only four-firm concentration ratio available for a paper and pulp industry subcategory to include some firms that did not produce the products (or use the process) generating the particular effluent stream being regulated.

Because no one inside or outside the agency knew with certainty the correct numbers for these industry characteristics, those numbers appearing in the economic analysis report or development documents were assumed to be the primary influence on agency decision making. Thus, the data used here reflect EPA perceptions of industry characteristics rather than the industries' true, but unknown, characteristics.

Because of gaps in EPA data on profitability, $ROR$ measures were constructed from the Internal Revenue Service's (IRS) enterprise data for 1972 (which classifies firms according to dominant product classes) and linked to EPA sectors according to their SIC codes. The definitions of industry used to construct aggregate profit rates do not perfectly match those used by EPA, however, so profitability results must be interpreted with caution.

Data on concentration ratios ($CR4$); numbers of small plants ($NSP$), defined according to the Bureau of the Census as a plant employing fewer than twenty employees; and number of plants ($NP$) were found more often in published EPA documents than were the data on profitability. Census or IRS sources provided an alternative source of data when the EPA documents did not contain concentration ratios. Actually, as discussed in appendix 5-B, the presence of conflicting estimates for these factors within published EPA documents was more of a problem than the unavailability of data.

The EPA documents estimated for each subcategory the number of plants "at risk" (denoted by $NPAR$); that is, the number of plants expected to incur costs to meet the effluent guideline standards. This number was less than the number of plants because some plants often discharged their wastes to municipal treatment systems (and were therefore exempt) and others were already meeting the standards because of process or waste-recovery considerations or compliance with previous federal or state regulations. The number of plants at risk was an important variable to EPA because it was used, along with model plant costs, to calculate the aggregate compliance costs for plants in the subcategory. A data series for $NPAR$ and $PPAR$ (the percentage of plants at risk) was constructed from earlier work (Gianessi, 1976) and, as a back-up, from the basic documentation of the BPT process. Both the Effluent Guidelines Division and the Economic Analysis Division estimated the number of plants the regulations would affect. Because these estimates often differed dramatically, however, a choice had to be made between them (of which more is said in appendix 5-B).

Testing the hypothesis relating the stringency of standards to the relative heights of the total abatement cost functions for different firms (through the parameter $r$) required measures of the relative heights of the total abatement cost functions for firms in different subcategories. Two different, but related, industry characteristics served as such a measure—waste-water flow ($FLOW$) and raw waste load ($RAW$).

The literature on industrial waste treatment shows that, for a constant raw waste load, higher wastewater flows raise the abatement costs of achieving any desired level of BOD or TSS pollution removal, or, equivalently, any given standard denominated in pounds of pollution per dollar of sales.[12] Holding $RAW$ constant, higher flows reduce pollution concentrations, however, implying that TSS standards based on units of concentration are less costly to meet as flow increases. Thus, for the three measures of the stringency of standards used here, $FLOW$ provides a direct proxy for the relative heights of the abatement cost functions.

The engineering literature suggests that $RAW$ can also serve as a proxy for the relative heights of abatement cost functions in regressions explaining variation in TSS standards (measured as a concentration or

[12] See Van Note (1975).

discharge per dollar of sales). Holding *FLOW* constant, a higher *RAW* for TSS requires more expensive treatment to meet any given standard.

The relationship between the raw waste load of BOD and the height of the abatement cost function is more complex. From an engineering point of view, abatement costs are roughly proportional to the percentage BOD removal for any quantity of raw waste being treated. Thus, percentage removal is a good proxy for the cost of BOD removal and should be unrelated to the raw BOD waste load. In contrast, the stringency of BOD standards measured in terms of pounds discharged per dollar of sales should be directly related to the size of *RAW* because a higher initial loading requires a greater percentage removal of BOD to meet any given discharge standard.

Both raw waste and flow data for each subcategory were taken from Gianessi (1976) and the draft development documents. These variables were normalized by sales by direct dischargers. This normalization makes raw waste and flow data comparable across subcategories and industries by correcting for subcategory size differences. In our specification, only the standards recommended by the contractor, not changes in the standards, are related to these variables. Raw waste and flow information should have been central to the contractor's analysis but should not have influenced changes in the standards unless the contractor made an error or new information on raw waste loads and flows was found.

*Industry Impacts.* Five industry impact variables were used—levels of unemployment (*UNEMP*), plant closings (*CLOSE*), percentage price increase (*PI*), and investment (*INV*) and annual (*ANNUAL*) costs projected to result from the proposed rules.[13] Data on each of these variables were gathered from the economic analysis reports. Both the projected number of plant closings and projected unemployment were normalized by the number of plants the standards affected.[14] The investment and annual compliance costs were normalized by sales by the plants in the subcategory that were direct dischargers. If these costs are thought of as the product of a cost coefficient (for compliance or investment cost) in cents per pound of production multiplied by production, then dividing by sales yields annual compliance or investment costs (cents per pound of production) as a percentage of selling price (assuming prices remain constant after the regulation goes into effect).

---

[13] Note that in projecting percentage increase in product prices caused by a standard EPA assumed, unrealistically, that 100 percent of the compliance costs would be passed through to consumers in the form of higher prices.

[14] The percentage unemployed could not be used because employment data at the subcategory level of detail often were not available.

Two industry impacts, *ANNUAL* and *INV*, provide alternative measures of the relative heights of the abatement cost functions of plants in different subcategories. The two industry characteristics, *FLOW* and *RAW*, also serve as such measures. *ANNUAL* and *INV* measure the heights of two related cost curves at the standards suggested by contractors (that is, for individual points on the curves rather than for the relative positions of the entire cost curves). They could be misleading measures of the relative heights of total abatement cost curves if the total cost curves are not systematically related to the annual and investment cost curves, or if the curves cross. Because the *RAW* and *FLOW* variables are more directly related to the heights of the total abatement cost curves, they were used to supplement the more obvious cost proxies of *ANNUAL* and *INV*.

*Internal Variables.* Many internal, information-based variables also influence the regulatory process. Data from the BPT process were available for only four of them: (1) turnover (that is, replacement) of the technical project officer for an industry during the course of the rule-making procedure (denoted by *TURN*); (2) the chronological order in which the standards for industries were promulgated (*ORDER*); (3) the quality of the data; and (4) the outcomes of earlier stages of the process, which can also be used to explain later outcomes. Only the last two require further explanation.

The quality of information developed by the agency was hypothesized to be directly related to the stringency of the standards. The development documents (including the contractor reports) and economic analysis reports were, by all accounts, the key documents in the rulemaking process, so the quality of the information contained in them should be most relevant to this hypothesis. Because the two sets of documents contained very different types of information and were developed by different groups, document quality could differ between the two for the same industry. Thus, two measures of quality—one for the development documents and one for the economic analysis reports—were devised.

A relation between quality and stringency exists because an agency would have a difficult time defending its decision against industry or other critics if it could be shown that the agency had relied on incomplete, inconsistent, or inaccurate data. Thus, a high-quality development document is defined as one in which critical information for setting standards is present and consistent throughout the document. In the development documents, this critical information concerns data used to transform the published standards into measures of relative stringency.

These data include raw waste load, flows, and current emission levels. Previous attempts to extract this information (Patterson, 1976) and our experience were used to make this judgment. The quality of the economic analysis reports was judged according to whether the method used to estimate incremental investment and annual costs for all plants in the subcategory could be discerned and duplicated. We relied primarily on previous work by Gianessi (1976) for this information. One dummy variable for the quality of the development documents (*DDQUAL*) and one for the quality of the economic analysis reports (*EAQUAL*) were constructed for each industry.[15]

The number of pollutants that were regulated (*POLL*) was used as an independent variable measuring the complexity of the waste stream and, therefore, the regulatory process. With industry's information advantage and the technical difficulties associated with treating many pollutants, values of *POLL* should be inversely related to the quality of information available to support the regulation. Note, however, that *POLL* is an outcome variable, that is, a decision the agency made during the contractor stage of the BPT rulemaking process, although one virtually unchanged throughout the process and statistically unrelated to other dependent or independent variables.

The hypothesis concerning the dynamics of the rulemaking process and its expression in the recursive econometric model suggested six additional variables. These variables explain outcomes of later stages of the rulemaking process based on the outcomes of earlier stages. The contractor standard may be related to the change in the standard from stages C to P; and this outcome may in turn, along with the proposed standard, be related to the change in the standard from stages P to R. For BOD and TSS, six additional variables are used to explain the various outcomes (all of which are defined in table 5-2): *BODC, TSSC, BODP, TSSP, CPB,* and *CPT.*

Table 5-4 lists the explanatory variables used in the analysis, along with the hypotheses they are used to test. Summary information on each variable appears in table 5-5. Table 5-6 contains the zero-order correlation coefficient matrix for all the explanatory variables. Note from

[15] Although finer judgments about the relative quality of documents are possible, we believed that the results would be less ambiguous if documents were classified as either of high or of low quality. As the key documents were updated only in very minor ways during the process, document quality did not vary within a rulemaking process for a subcategory. Neither did quality vary over subcategories in a given industry, because standards or economic effects for all subcategories were analyzed at the same time by the same contractor or project officer, using identical methodologies.

TABLE 5-4.  EXPLANATORY VARIABLES

| Variable | Description | Hypothesis number[a] |
|---|---|---|
| CR4 | Four-firm concentration ratio (1972) | 6 |
| ROR | Rate of return on equity for industry (1972) | 5 |
| TAB | Trade association budget (thousands of dollars, 1972) | 7 |
| UNEMP | Projected unemployment from meeting the proposed standards as a percentage of plants at risk | 3 |
| CLOSE | Projected closings from meeting the proposed standards as a proportion of plants at risk | 3 |
| PI | Projected percentage product price change needed to meet the proposed standards | 3 |
| INV | Projected investment costs of meeting the proposed standards per dollar of sales for plants at risk | 8 |
| ANNUAL | Projected annualized costs of meeting the proposed standards per dollar of sales for plants at risk | 8 |
| NC | The number of comments appearing in the *Federal Register* that argue for a weakening of the standards. A *B* after *NC* refers to comments on BOD standard, a *T* to comments on TSS, and a *G* to general comments. These designations may be followed by a *P* for comments received in the first comment period, an *R* for comments received in the second comment period, or a *CR* for the sum of comments received during both periods. Four separate comment variables are used in different regression equations reported in chapter 6: *NCBGP*, *NCBGCR*, *NCTGP*, and *NCTGCR*. | 2 |
| NPAR | Number of plants at risk (that is, the number of plants that will incur compliance costs) | 1 |
| PPAR | Proportion of plants at risk | 4 |
| NSP | Number of small plants (usually defined as plants with fewer than twenty employees) | 9 |
| PSP | Percentage of small plants | 9 |
| RAWB | The ratio of BOD raw waste load (in kkg) to sales (in millions of 1972 dollars) of direct dischargers | 10 |
| RAWT | The ratio of TSS raw waste load (in kkg) to sales (in millions of 1972 dollars) of direct dischargers | 10 |
| FLOW | The ratio of waste-water flow (in millions of liters) to sales of direct dischargers | 10 |
| DDQUAL | The quality of the draft Development Document (dummy variable) | 11 |
| EAQUAL | The quality of the Economic Analysis report (dummy variable) | 11 |
| TURN | Turnover in EGD project officers for a subcategory during the regulatory process (dummy variable) | 13 |
| ORDER | The order in time in which an industry's standards were promulgated | 14 |
| POLL | The number of pollutants for which proposed standards were written | 11 |

[a] See chapter 4 for a statement of each hypothesis.

TABLE 5-5. DESCRIPTIVE STATISTICS OF EXPLANATORY VARIABLES

| Variable | $N^a$ | Mean | Standard deviation | Minimum | Maximum |
|---|---|---|---|---|---|
| CR4 | 105 | 40.96 | 21.98 | 15.00 | 100.00 |
| ROR | 105 | 6.70 | 1.81 | 3.90 | 10.10 |
| TAB | 106 | 2,712.58 | 1,805.65 | 0 | 10,000.00 |
| UNEMP | 106 | 6.30 | 13.10 | 0 | 65.38 |
| CLOSE | 106 | 0.12 | 0.24 | 0 | 1.00 |
| PI | 106 | 1.35 | 2.05 | 0 | 8.00 |
| INV | 105 | 0.10 | 0.21 | 0 | 1.43 |
| ANNUAL | 105 | 0.04 | 0.08 | 0 | 0.44 |
| NCBGCR | 106 | 10.44 | 6.25 | 1.00 | 24.00 |
| NCBGP | 106 | 3.40 | 1.98 | 0 | 12.00 |
| NCBGR | 106 | 6.31 | 4.27 | 0 | 20.00 |
| NCTGCR | 106 | 10.37 | 6.34 | 0 | 27.00 |
| NCTGP | 106 | 3.48 | 1.85 | 0 | 11.00 |
| NCTGR | 106 | 6.32 | 4.64 | 0 | 23.00 |
| NPAR | 106 | 51.45 | 126.26 | 0 | 830.00 |
| PPAR | 105 | 0.40 | 0.33 | 0 | 1.00 |
| PSP | 105 | 46.44 | 25.98 | 0 | 100.00 |
| RAWB | 69 | 88.44 | 134.14 | 0 | 643.68 |
| RAWT | 101 | 1,101.16 | 3,980.82 | 0 | 36,261.30 |
| FLOW | 101 | 254.56 | 582.84 | 0 | 4,048.19 |
| DDQUAL | 106 | 0.44 | 0.50 | 0 | 1.00 |
| EAQUAL | 106 | 0.26 | 0.44 | 0 | 1.00 |
| TURN | 106 | 0.25 | 0.43 | 0 | 1.00 |
| ORDER | 106 | 3.79 | 2.41 | 1.00 | 11.00 |
| POLL | 106 | 3.97 | 1.80 | 0 | 12.00 |

a Number of observations.

table 5-6 that correlation coefficients are low between all independent variable pairs except UNEMP-CLOSE ($r = .70$) and NCBGCR-TURN ($r = .54$).

## Data Problems

Some of the practical problems of collecting and adjusting data for use in a statistical analysis of rulemaking include unavailable data, data omissions and inconsistencies in the documents, the measurement units problem, and the aggregation problem. Appendix 5-B treats these problems in more detail and provides the rules of thumb that were used for resolving them. The data problems addressed in appendix 5-B are generally less difficult than the methodological issues discussed in this chapter and are typical of the types of data problems confronted in many other kinds of empirical research.

TABLE 5-6. CORRELATION COEFFICIENT MATRIX OF EXPLANATORY VARIABLES[a]

| Variables | CR4 | TAB | ROR | PSP | ORDER | POLL | NPAR | EAQUAL | DDQUAL | TSSC[b] | BODC[b] |
|---|---|---|---|---|---|---|---|---|---|---|---|
| CR4 | — | | | | | | | | | | |
| TAB | -.31 | — | | | | | | | | | |
| ROR | .27 | -.24 | — | | | | | | | | |
| PSP | -.26 | .04 | .19 | — | | | | | | | |
| ORDER | .11 | -.18 | -.12 | -.18 | — | | | | | | |
| POLL | -.03 | -.43 | .08 | .04 | .13 | — | | | | | |
| NPAR | -.23 | -.01 | .18 | .17 | .25 | .11 | — | | | | |
| EAQUAL | .17 | .41 | .32 | -.12 | .30 | -.18 | .32 | — | | | |
| DDQUAL | -.08 | -.11 | .05 | .19 | -.16 | -.34 | -.16 | .10 | — | | |
| TSSC | .24 | .15 | -.11 | -.16 | .16 | -.12 | -.05 | -.04 | .1 | — | — |
| BODC | -.60 | -.18 | -.14 | .03 | -.12 | .28 | .15 | .00 | -.36 | — | — |
| BCP | -.18 | -.24 | .06 | .06 | -.01 | .18 | -.09 | .00 | -.24 | — | -.50 |
| TCP | -.20 | -.20 | .16 | -.02 | .08 | .27 | -.04 | -.06 | -.31 | .26 | — |
| TURN | -.21 | -.20 | -.08 | -.14 | .07 | .28 | -.11 | .20 | -.06 | -.06 | .08 |
| PPAP | .22 | -.02 | -.11 | -.26 | -.11 | -.13 | -.12 | -.10 | -.15 | .24 | -.26 |
| UNEMP | .02 | -.08 | -.17 | .00 | .26 | -.01 | .10 | -.13 | -.02 | -.04 | .03 |
| INV | -.08 | .19 | -.01 | .06 | .03 | -.18 | -.08 | -.16 | .22 | .04 | -.17 |
| CLOSE | .00 | -.02 | -.22 | .18 | .13 | .13 | -.06 | -.01 | -.09 | -.03 | .29 |
| PI | .03 | .12 | -.07 | -.25 | -.04 | -.13 | .09 | -.05 | -.04 | .00 | -.15 |
| NCBGCR | -.34 | -.05 | .03 | .00 | -.15 | -.01 | -.09 | -.20 | -.20 | .00 | .28 |
| RAWB | -.04 | .34 | -.26 | -.20 | .09 | -.20 | -.21 | -.22 | .14 | — | .48 |
| RAWT | .00 | .21 | .03 | -.13 | -.14 | -.13 | -.09 | -.16 | .28 | -.40 | — |
| FLOW | .02 | .16 | -.06 | -.06 | -.03 | -.14 | -.11 | -.18 | .11 | -.02 | -.16 |

[a] Pearson product-moment correlation coefficients.

[b] The following example explains the interpretation of correlation coefficients in these columns: take the intersection of the column headed by $\frac{TSSC}{BODC}$ with the row headed by TURN. Then $-.06$ is the correlation coefficient for TSSC and TURN, while $.08$ is the correlation coefficient for BOD and TURN.

| BCP[b]<br>TCP | TURN | PPAR | UNEMP | INV | CLOSE | PI | NCBGCR | RAWB[b]<br>RAWT | FLOW | Variables |
|---|---|---|---|---|---|---|---|---|---|---|
| | | | | | | | | | | *CR4* |
| | | | | | | | | | | *TAB* |
| | | | | | | | | | | *ROR* |
| | | | | | | | | | | *PSP* |
| | | | | | | | | | | *ORDER* |
| | | | | | | | | | | *POLL* |
| | | | | | | | | | | *NPAR* |
| | | | | | | | | | | *EAQUAL* |
| | | | | | | | | | | *DDQUAL* |
| | | | | | | | | | | *TSSC* |
| | | | | | | | | | | *BODC* |
| — | | | | | | | | | | *BCP* |
| — | | | | | | | | | | *TCP* |
| .26<br>.55 | — | | | | | | | | | *TURN* |
| −.12<br>−.07 | .05 | — | | | | | | | | *PPAP* |
| .15<br>.24 | .11 | .08 | — | | | | | | | *UNEMP* |
| −.17<br>−.12 | .00 | −.04 | .06 | — | | | | | | *INV* |
| .18<br>.29 | .29 | .18 | .70 | .02 | — | | | | | *CLOSE* |
| −.26<br>.00 | .31 | .06 | .08 | .37 | .09 | — | | | | *PI* |
| .29<br>.47 | .54 | .16 | −.11 | .13 | .11 | .25 | — | | | *NCBGCR* |
| .01<br>— | .23 | .37 | −.10 | .42 | −.06 | .21 | .26 | — | | *RAWB* |
| —<br>−.09 | −.14 | −.11 | −.10 | .01 | −.11 | −.11 | .03 | — | | *RAWT* |
| −.09<br>−.09 | −.01 | −.03 | .02 | .10 | −.01 | .02 | .04 | .49<br>.38 | — | *FLOW* |

## SUMMARY

Developing the data base proved to be one of the most time-consuming and difficult phases of applying the revealed preference approach to the BPT process. Merely expressing the bureaucratic outcomes in a way that allows comparisons of stringency across industries was a major undertaking. Because of this difficulty, two different units were used to measure the level of the stringency of the standards. Measures for changes in stringency over a given regulatory process were much easier to develop. Finding and reconciling data to match the appropriate aggregated form of the industrial subcategories required the assumption that information critical to the decision-making process was included in the major source documents and that the EPA perceptions about the size of certain parameters, rather than their true size, were the important factors in explaining bureaucratic outcomes. The data base that resulted from this effort permits tests of many of chapter 4's hypotheses about the stringency of standards, but not those on categorization.

## APPENDIX 5-A
*The Federal Regulations Data Base*

The computerized Federal Regulations Data Base (FRDB) consists of a standards file and an economic and technical information file. The standards file contains effluent limitations on ninety-six pollutants covering the forty-nine Group One industries regulated by EPA. Each regulation is tracked through three stages of the regulatory process. The economic and technical information file contains data on expected economic impacts of the regulations, industry characteristics, and other factors related to the regulatory process.

### The Standards File

Data in the standards file are organized by industry subcategory. Each subcategory is designated by name, the basis of subcategorization, and by industry-subcategory identification number. The units in which the regulations were written, the thirty-day average and one-day maximum standards for each pollutant, and notes for special cases appear with each subcategory. Table 5A-1 lists the thirty-seven bases used for subcategorization; the most prevalent of these were goods produced and production process. Table 5A-2 lists the ninety-six pollutants for which regulations were issued, though no single subcategory or industry had all of them in its waste stream. BOD5, TSS, and pH were the pollutants most often regulated.

TABLE 5A-1.  SUBCATEGORY TYPES

| | |
|---|---|
| Age | Intake water quality |
| Location | Laboratory |
| Product | Land availability |
| Process | Maintenance wastes |
| Size | Miscellaneous streams |
| Type of plant | Process residues |
| Air pollution controls | Process water |
| Ash handling | Radioactive |
| Ballast | Rainfall/runoff |
| Barometric condenser | Raw materials storage |
| Blowdown | Raw materials type |
| Cleaning | Sanitary |
| Chemical wastes | Transformer fluid |
| Cooling water | Unregulated wastes |
| Contaminated non-process water | Waste concentration |
| Construction | Waste ponds |
| Drainage | Waste volume |
| Fuel type | Water treatment |
| Intake backwash | |

116

TABLE 5A-2. REGULATED POLLUTANTS

| | |
|---|---|
| Acidity | Other metals |
| Alkalinity | Dissolved other metals |
| Aluminum | Mercury |
| Aluminum ($Al_2O_3$) | Dissolved mercury |
| Dissolved aluminum | Nickel |
| Arsenic | Dissolved nickel |
| Barium | Nitrate |
| Cadmium | Nitrate/nitrite |
| Calcium | Organic nitrogen |
| Total organic carbon (TOC) | Total Kjeldahl nitrogen |
| Chemical oxygen demand (COD) | Oil |
| Chloride | Oil and grease |
| Free available chlorine | Organics |
| Chromium ($Cr_2O_3$) | Osmium |
| Dissolved chromium | Oxidizing agents |
| Hexavalent chromium | Biological oxygen demand (BOD) |
| Total chromium | Chemical oxygen demand (COD) |
| Cobalt | Total oxygen demand (TOD) |
| Fecal coliforms | Palladium |
| Total coliforms | PCB |
| Color | Phenol |
| Copper | Phosphate |
| Dissolved copper | Ortho phosphate |
| Dissolved cyanide | Elemental phosphorus |
| Oxydizable cyanide | Total phosphorus |
| Cyanide-A | Platinum |
| Total cyanide | Potassium |
| Debris | pH |
| Fluoride | Rhodium |
| Fish solids | Ruthenium |
| Gold | Selenium |
| Total hardness | Silicon |
| Heat | Silver |
| Hydrogen sulfide | Silver ion |
| Iridium | Sodium |
| Iron | Settleable solids |
| Iron ($Fe_2O_3$) | Total dissolved solids (TDS) |
| Dissolved iron | Total suspended solids (TSS) |
| Ferrous iron | Total volatile solids |
| Soluble iron | Total solids |
| Lead | Sulfate |
| Dissolved lead | Sulfide |
| Magnesium | Surfactants |
| Manganese | Temperature |
| Dissolved manganese | Tin |
| Heavy metals and toxics | Vanadium ($V_2O_5$) |
| Any heavy metals | Zinc |
| All heavy metals | All pollutants |

*Note:* Regulated pollutants are specific pollutants (or pollutant categories) for which regulations were issued for at least one subcategory during the process.

Each subcategory entry includes all the discharge standards developed throughout the regulatory process. Consequently, a level code associated with each set of standards identifies the regulatory stage (C, P, or R) at which the standards were written. Since the subcategorization was often changed between stages, a compatibility code shows whether a subcategory was newly created at a given level, was carried through unchanged from the previous level, or was linked in more complex ways to a subcategory created at an earlier stage; for instance, a subcategory at the proposed stage could be split into three subcategories at promulgation, each with its own set of standards.

## Economic and Technical Information File

From the population of forty-nine Group I industries (a total of more than 600 EPA-defined subcategories at promulgation), a sample of twenty-three industries was selected for statistical analysis. Since data on economic characteristics or expected impacts were often unavailable at the level of EPA-defined subcategorization, aggregations of the 268 sample subcategories resulted in a final total of 106 subcategory observations (see table 5-1). Variables describing the stringency of regulations were created directly from the standards file and from information contained in the development documents [DD].[16] Other variables were drawn from a variety of sources. In addition to [DD], these sources include

[EA]     Economic analysis reports commissioned by EPA to support the development of proposed and promulgated regulations

[TD]     Working group economic impact assessments based on [EA] and [DD], summarized in "TAB D" of EPA Action Memoranda

[G]      *Estimates of National Water Pollutant Discharges by Polluting Sector: 1972*, Appendix B1, L. Gianessi, 1978

[CI]     *The Cost to Industries of Meeting the 1977 Provisions of the Water Pollution Control Act Amendments of 1972*, L. Gianessi and H. Peskin, 1976

[IRS]    *Statistics of Income, 1972: Corporate Income Tax Returns*, Internal Revenue Service, 1972

[SS]     *1972 Census of Manufactures, Vol 1: Subject and Special*

---

[16] Abbreviated source information appears in brackets [  ].

118 RULES IN THE MAKING

Statistics, Bureau of the Census, U.S. Department of Commerce, 1976

[NTPA] National Trade and Professional Associations of the United States (Washington, D.C., Columbia Books, Inc., 1972, 1973, 1974)

[P] Directory of Federal and State Water Pollution Control Standards, J. Patterson, Illinois Institute for Environmental Quality, State of Illinois Document #77/06 (Springfield, IIEQ, 1976)

[FRP] Federal Register, Interim and Proposed Rules

[FRR] Federal Register, Final Rules and Regulations

[EPA] Internal EPA documents

[SF] RFF Standards File

[I] Interviews with EPA project officers and other key EPA personnel

[RFF] Evaluations by RFF personnel of the ease of data extraction and internal consistency of EPA documents

The following subsections briefly describe the variables contained in the file.

*Variables Describing Stringency.* These variables include percentage changes in standards over the contractor, proposed, and promulgated stages of the process; the promulgated BOD5 and TSS standards; the number of pollutants regulated at promulgation; and the number of promulgated subcategories, by industry. Percentage changes in standards were computed directly from [SF], as were the number of subcategories into which an industry was divided and the number of regulated pollutants within a given industry or subcategory. In order to facilitate cross-industry comparisons, the promulgated BOD and TSS standards were converted from the units in which they were written (usually kg pollutant per kkg production) to percentage removal from raw waste levels (for BOD5) and milligrams per liter of effluent flow (for TSS). If subcategory aggregations were necessary, percentage change values were weight-averaged on the basis of subcategory production shares; subcategory raw waste levels and flow values were used to weight-average the percentage removal and milligrams per liter standards, respectively.[17] By convention, the calculated percentage change from zero

[17] These transformations were aided by work done by Patterson (1976).

discharge to a specific limitation was expressed as the change in percentage removal from the corresponding raw waste level.[18]

The ten variables describing stringency are[19]

BCR    Percentage change in thirty-day average BOD5 effluent limitation, contractor to promulgated [SF]

BCP    Percentage change in thirty-day average BOD5 effluent limitation, contractor to proposed [SF]

BPR    Percentage change in thirty-day average BOD5 effluent limitation, proposed to promulgated [SF]

TCR    Percentage change in thirty-day average TSS effluent limitation, contractor to promulgated [SF]

TCP    Percentage change in thirty-day average TSS effluent limitation, contractor to proposed [SF]

TPR    Percentage change in thirty-day average TSS effluent limitation, proposed to promulgated [SF]

BODC   Thirty-day average BOD5 effluent limitation, expressed as percentage removal from average or model plant raw waste load (as appropriate), contractor stage [SF,DD]

TSSC   Thirty-day average TSS effluent limitation, expressed in milligrams per liter of average or model plant waste-water flow (as appropriate), contractor stage [SF, DD, P, G]

NSCR   Number of subcategories for which standards are promulgated [SF]

POLL   Number of pollutants for which standards promulgated [SF]

*Variables Describing Effects.*    Effects variables include measures of expected effects on industries assuming that the proposed standards are promulgated without change. In [CI] and [G], EPA's Economic Analyses and Development Documents were analyzed and estimates of costs and other variables made. These estimates were based only on information available during preparation of the proposed rules. When they matched our subcategorization, the [CI] and [G] estimates were used.

---

[18] For example, consider a hypothetical BOD5 waste load of 5.0 kg/kkg. If the standard changes from zero discharge (or 100.0 percent removal) to 2.5 kg/kkg (or 50.0 percent removal), the calculated percentage change will be −50.0 percent. In effect, this represents a "normalized" percentage change calculation (adjusted to reflect the initial raw waste level).

[19] Percent changes are actually calculated as average changes between levels so as to minimize the units bias. See the discussion in chapter 5.

When [CI] and [G] were not usable, [TD] was used as second best and [EA] was used as a default.

The effects variables are

ANNUAL       Expected annual compliance costs, including operation and maintenance and investment (amortized in 1972 dollars), of meeting proposed standards [I, G, TD, EA]

INV          Expected investment costs (1972 dollars) of meeting proposed standards (note that plant depreciation rates are not constant across industries) [I, G, TD, EA]

CLOSE        Number of predicted plant closures proposed standards would cause [TD, EA]

ICR          Dummy variable for whether the proposed standard would significantly affect the subcategory's import competitiveness

NPAR         Number of plants expected to be directly affected by the proposed regulations; includes only plants that discharge directly into waterways or holding ponds and do not have the recommended BPT technology in place [I, G, EA, DD]

PRICE        Expected percentage increases in average selling prices of goods attributable to the costs of meeting the proposed standards [TD, EA]

UNEMP        Number of jobs predicted to be lost as a result of the proposed standards [TD, EA]

*Variables Describing Industry Activities.*   These variables describe industry structure and performance.

CR4          Four-firm concentration ratios applied to each industrial subcategory. Where SIC codes did not correspond directly to EPA-defined subcategories or industries, concentration ratios for each SIC included in the relevant subcategory or industry were weight-averaged according to SIC sales totals. *CR4* data in [EA] were given priority [G, SS, EA]

XCR4         The same as *CR4* but data in [SS] were given priority [G, SS, EA]

FLOW         Millions of liters of water per year discharged by plants in the appropriate subcategory [CI, G, DD]

NPLANT       The total number of plants in the relevant subcategory [G, TD, EA, DD]

| | |
|---|---|
| *PBPT* | The percentage of discharging plants with BPT technology already in place [I, G, TD, EA, DD] |
| *PROD* | Production in direct-discharging plants, in units compatible with the regulations (generally kkg of production) [I, TD, EA, DD] |
| *PSP* | The percentage of establishments in the subcategory with fewer than 20 employees; using the relationship between SIC codes and EPA-defined subcategories in [G], *PSP* is defined as the ratio of the number of plants with fewer than 20 employees to the total number of plants in the appropriate SIC [SS, G] |
| *RAWB* | Yearly amount of subcategory or industry BOD5 raw discharge; found by multiplying the raw waste coefficient for the average plant by yearly production in direct-discharging plants, or taken directly from [G] [CI, G, DDI] |
| *RAWT* | Same as *RAWB* for TSS |
| *ROR* | Pretax rate of return on assets (ratio of 1972 pretax income to total 1972 assets). SIC codes were matched with IRS standard enterprise classification codes by subcategory (or by industry) and then applied to each subcategory[20] [IRS] |
| *SALES* | Total sales for the subcategory in millions of 1972 dollars [TD, EA, DD] |
| *TAB* | Trade association yearly budget applied to each subcategory, averaged over 1972–1974 for each industry. Trade associations mentioned in [FR] or active in BPT litigation were surveyed, and the budgets of "significant" associations (those taking the lead in industry coordination, litigation, and overall strategy) were tabulated. Midpoints were taken when a budget range was given [NTPA, TAS] |

*Other Variables.*

| | |
|---|---|
| *CF1-CD10* | Dummy variables for specific contractors (eleven in all) hired by EPA to draft the development document [EPA] |
| *CONEX* | Dummy variable for the contractor's prior experience with waste management problems of the industry [I] |

---

[20] With multiproduct plants subject to guidelines across subcategory or industry classifications, this "matching" at high levels of aggregation, though not exact, provides a reasonable estimate of effective rates of return (see table 5A-3).

TABLE 5A-3.   CORRESPONDENCE BETWEEN STANDARD INDUSTRIAL
CLASSIFICATION AND STANDARD ENTERPRISES CLASSIFICATION
CODES

| Industry | SIC | SEC |
|---|---|---|
| Leather | 3111 | 31 |
| Textiles | 223, 225, 227 | 22 |
| Builders paper | 2661 | 261, 262, 263, 266 |
| Cement | 3241 | 32 |
| Phosphates II | 2874, 2819 | 286, 287, 289, 281, 282 |
| Sugar I | 206 | 206 |
| Paper & pulp I | 2611, 2621, 2631 | 261, 262, 263, 266 |
| Timber I | 2435, 2436, 2499, 2491 | 24 |
| Glass II | 3221, 3229 | 321, 322, 323 |
| Meats I | 2011 | 201 |
| Meats II | 2013 | 201 |
| Glass I | 3211, 3231 | 321, 322, 323 |
| Timber II | 2421, 2492, 2661 | 24 |
| Phosphates I | 2819 | 281, 282 |
| Sugar II | 2061 | 206 |
| Organics I | 2865, 2869, 2861 | 286, 287, 289 |
| Organics II | 2865, 2869, 2861 | 286, 287, 289 |
| Inorganics I | 2812, 2819, 2834, 2873 | 281, 282 |
| Rubber I | 3011, 282 | 281, 282, 301, 302, 303, 304, 306 |
| Rubber II | 3021, 3041, 3293, 3069, 7534 | 301, 302, 303, 304, 306 |
| Poultry | 2016, 2017 | 201 |
| Fruits & vegetables I | 203 | 203 |
| Fruits & vegetables II | 203 | 203 |

Source: *Standard Industrial Classification Manual*, 1972, Office of Management and Budget; and
*Statistics of Income, 1972: Corporate Income Tax Returns*, Internal Revenue Service, 1972.

| | |
|---|---|
| CONQ | Dummy variable for the quality of the contractor's work as judged by the project officer [I] |
| TURN | Dummy variable indicating whether the EPA project officer changed during the BPT process [EPA] |
| DDQUAL | Characterization of the relative quality of EPA development documents, based on our ability to extract information from them [RFF] |
| DNPP | Dummy variable indicating whether limitations were proposed for pollutants other than BOD5, TSS, and pH [SF] |
| DNPR | Same as DNPP but for the promulgated regulations |
| EAQUAL | Dummy variable for the quality of [EA] derived from [G] a quality document is defined as one using a |

clear and consistent methodology in computing the costs of complying with the regulations; the contractor and project officer were given the opportunity to explain their methodology by [G] [RFF]

MEDIA         Dummy variable indicating whether high projected costs of pollution control to an industry received "significant" media attention during the BPT process [I]

NCBG          The number of comments pertaining to BOD5 published in [FR]; this variable is further differentiated according to level (that is, whether the comment came in response to the draft DD, the proposed DD, or both) and subcategory or industry applicability (that is, whether the comment applies to a specific subcategory or is directed toward the industry in general) [FRP, FRR]

NCTG          Same as NCBG for TSS

POLL          The number of pollutants regulated at the proposed stage [SF]

OPRES         Dummy variable indicating presence of significant pressure from the White House, OMB, any other federal agency, or the Congress to relax the standards [I]

ORDER         Chronological ranking of industry regulations based on date of promulgation [FRR]

PHR           Dummy variable for the phase of the industry (EPA regulated an industry in phases to leave more complex and/or less important industry segments for later in the process) [EPA]

POD1–POD10    Dummy variables identifying the EPA project officers (eleven in all) for each industry [EPA]

APPENDIX 5-B
*Data Problems Raised by the Methodology*

The application of the revealed preference approach to the BPT rule-making process posed a variety of practical problems of collecting and adjusting the data. These problems include unavailable data, data omissions and inconsistencies in the data collected from different sources, units of measurement, and data aggregation.

## Unavailable Data

Guided by only the general theories of rulemaking presented in chapter 3 and attempting to describe a complex process, we sought to accumulate information from all possible written and oral sources—from the official documents underlying the regulations to telephone logs, from interviews with EPA administrators to interviews with OMB staff. We originally hoped to find most of the key documents in one place. Unfortunately, EPA did not, and does not, have a central catalog, storage, or retrieval system.

The closest thing to such a system was EPA's "administrative record" for each industry, compiled in case the standards were challenged in court. This record has several drawbacks. First, it is voluminous: twenty-seven boxes for fruits and vegetables (phase II) alone. What is more, an index to the records exists only for industries for which the standards were litigated. In addition, access is difficult because only EPA personnel are permitted to go through the records, which must be taken out of storage. The record also appears to be incomplete. Many interoffice memos, such as the key "action memoranda" sent to top administrators by lower echelon staff, are generally not in the record. Equally distressing are the pervasive gaps found by checking the index of comments against the comment files held in the EPA library. Thus, because of the great expense associated with analyzing the records and the certainty that key documents would still be missing, we were forced to use more disparate, but more readily available, sources of information.

The EPA library provided public documents, such as contractor reports, development documents, and economic analysis reports for some of the industries. This collection had to be supplemented by project officers or other EPA officials. Other specialized libraries, such as the Water Resources Council library, also turned up a number of otherwise unavailable documents. Several industry-prepared documents in support of an alternative set of standards or in opposition to EPA's standards were discovered as part of our information-gathering effort and used as background for interviews with EPA project officers. In addition, Resources for the Future staff contributed to our document collection. Finally, the *Federal Register* provided much of the data on the standards and summaries of industry comments, which, in the absence of a reliable source for the comments themselves, we were forced to use.

We interviewed many people to supplement the written record, including project officers for each industry, former EPA officials, and to a lesser extent, trade association representatives. Although some ex-project officers had trouble remembering details about the interventions of parties in the rulemaking process or controversies that surrounded particular standards, others shared vivid recollections. The purpose of interviews with trade association staff was to elicit budget and employment information, not to probe into their communications as part of the BPT rulemaking process.

*Data Omissions and Inconsistencies*

Perhaps the most painstaking aspects of building the data base were filling in gaps in the data and reconciling different estimates for the same variables. As an example of the former problem, extracting from a given development document the raw waste and flow information necessary to compute the percentage-removal and milligram-per-liter dependent variables created many difficulties. The latter problem is illustrated by our reconciliations of differing estimates of a given plant characteristic that appeared in the economic analysis reports, in the development documents, and, occasionally, in industry documents.

The most common problem associated with extracting raw waste and waste-water flow data was that a development document would often give flow and raw waste data both for average plants and for model plants (usually exemplary ones). When this problem occurred, we used average plant values because the computation of percentage removal of a pollutant for a subcategory required knowing how much the actual effluent discharges of typical plants have been reduced by the standards. In the case of the textiles industry standards, the development document does not provide any raw waste data other than the exemplary plant values, so it was impossible to compute percentage removals. For waste-water flows, justifying the use of average plant values is slightly more difficult: once the standard is imposed, plant performance might match that of exemplary plants and those flows will be appropriate. Because the standard itself is not generally written in milligrams per liter, however, and plants are free to use any technology that enables them to meet the standard, there is no a priori reason to expect producers to change their flows to produce the concentrations experienced by model plants. We therefore used average plant data whenever they were available.

Another common problem was that two types of plant flow data sometimes appeared—one for plants under the current regulatory regime and one for plants meeting BPT regulations. In these cases, we used the values on which the standards were based.

An additional problem occurred when a development document provided flow values and the effluent concentration anticipated at the BPT level of discharge, but the concentration calculated using the official standard (in pounds of discharge per thousand pounds of product) turned out to be a different one. In these cases, we went through the documents carefully to find other flow values, for instance, for a single exemplary plant rather than for all plants sampled. If a flow could be found in the document that made the calculated concentration measure equal to (or approximate) the EPA measure, we used that flow. If such a flow could not be found, we used the calculated concentration.

Beyond these rather systematic inconsistencies in the development documents, an array of other problems cropped up irregularly. These problems arose when data were not presented in useful units (for example, if flows were given in per-unit terms without any production figures). In these situations, we used our judgment as to the most appropriate flow and raw waste load data.

Inconsistencies between the economic analysis report and the development document for a specific industry were a common feature of the written record. For most industries in our data base, the data were inconsistent for at least one plant characteristic. Discrepancies arose particularly in the number of plants in an EPA-defined industry or subcategory and the number of plants already attaining various degrees of pollution abatement.

An example drawn from EPA and industry documents for the meat packing industry illustrates the problem and how it was addressed.[21] According to the Department of Agriculture's *Livestock Slaughter, Annual Survey (1972),* there were 5,991 meat slaughtering plants in the United States. Of these plants, 84 were "large," 309 were "medium," and the rest (under 200 million pounds annual live weight killed (LWK)) were "small." The EPA contractor for this industry, North Star Research Institute, estimated that 5,200 of the small plants were "locker" plants that slaughter and process meat. The remaining small plants were assumed to produce 1 to 25 million pounds LWK per year. Using a survey of thirty-three large and fifty-three medium plants, North Star divided these small plants into four EPA subcategories. In addition, it assumed that the small, nonlocker plants were equally divided between two of the subcategories. (Table 5B-1 indicates the industry make-up by EPA subcategory and size.) Because North Star recommended that locker plants be exempted from the standards, no subcategory was created for these plants.

---

[21] Much of the work reconciling data was performed as part of an earlier project and is reported in Gianessi (1978).

TABLE 5B-1.   MEAT PACKING PLANTS BY SIZE AND
INDUSTRY CATEGORY

| Type of plant | Large | Medium | Small | Total |
|---|---|---|---|---|
| Simple slaughterhouses | 2 | 103 | 200 | 305 |
| Complex slaughterhouses | 34 | 34 | — | 68 |
| Low process packinghouses | 30 | 86 | 200 | 316 |
| High process packinghouses | 18 | 86 | — | 104 |
| Locker plants | 0 | 0 | 5,200 | 5,200 |
| Total | 84 | 309 | 5,600 | 5,993 |

*Source:* Based on assumptions of North Star Research Institute, 1973.

Table 5B-2 shows how the economic analysis report divided the industry. Because the smallest plant mentioned in the report slaughters 2 million pounds LWK a year, the "small" categories in the two tables are not comparable. Also, the economic analysis report estimates show twenty-three more large plants and seventy-six more medium plants than North Star. We made some attempt to reconcile these estimates to determine if a better breakdown of plants would emerge, but we found that the economic analysis report incorrectly calculated the number of small and medium plants. Therefore, North Star's estimates were entered into the data base.

This example highlights a number of reasons for the discrepancies between estimates. First, because in many cases the two groups relied on different data sources for their analyses, differences in their results are not surprising. The Economic Analysis Division (EAD) contractors usually relied on economic publications, such as those of the Bureau of the Census, which use SIC codes; the Effluent Guidelines Division (EGD) contractors relied on special surveys, sampling, or industry data. The assumptions and definitions behind these data bases sometimes varied widely.

Second, because EGD's particular definitions of subcategories often lacked a counterpart in any existing data sources, the contractors found it necessary to make a large number of often arbitrary assumptions to

TABLE 5B-2.   ESTIMATED NUMBER OF MEAT PACKING AND
SLAUGHTER PLANTS

| Type of plant | Large | Medium | Small | Total |
|---|---|---|---|---|
| Meat packinghouses | 85 | 305 | 750 | 1,140 |
| Slaughterhouses | 22 | 80 | 178 | 280 |
| Total | 107 | 385 | 928 | 1,420 |

*Source:* EPA Economic Analysis Report, 1973.

aggregate or disaggregate the available data into the proper subcategories. The North Star assumption about the allocation of activities of small plants between slaughterhouses and packing houses is one such assumption.

Although we created a file-folder full of documentation on each industry in the process of developing a "best" number of plants, several aspects of our process are still troublesome. First, we could not always apply a consistent definition of "best." We adopted the rule of thumb that the information generated by the ultimate decision-making office, usually EGD, should be given priority. The EAD played a role in many decisions, however, and if that group could have demonstrated that EGD data were faulty, the EAD analyses would have carried more weight in setting standards. Thus, it was easy to choose between data sources when EGD data were judged superior. But when EAD data were judged to be more credible, we generally chose to use those data.

Second, we failed to make use of the discrepancies between estimates for the same factor, even though such discrepancies could reflect the degree of uncertainty held about any piece of information or perhaps the bargaining strength of particular groups (if some judgment about credibility could have been attached to each estimate). The lack of a theory on how such discrepancies might affect decisions is the primary reason for this omission. In addition, the data development problems would have been greatly magnified had the origin of each "best" data element, as well as the origin and value for each rejected data estimate, been included in the data base. Nevertheless, this potentially useful information is not irretrievably lost and can be included in future work.

*The Measurement Units Problem*

The measurement units problem comprises a host of minor, but meddlesome, problems, including (1) finding units that allow comparisons among attributes of subcategories in the same industries and in different industries, (2) converting data expressed in one measure to other more useful measures, and (3) converting data drawn from many different years to a common base year. The first measurement units problem mainly involved transforming data to make the dependent variables comparable. Raw waste loads per unit of production and the effluent standard (also in per unit of production) were used to compute required percentage removals; wastewater flow per unit of product and the effluent standards were used to calculate allowable concentrations; unit price or sales volume was used to convert the effluent standard to an emission per dollar of sales basis; and changes in the standard during the process were expressed in percentage terms. Key transformations of a number of independent variables were also made—such as dividing raw waste loads, flows, annual costs, and investment costs by sales of direct dischargers—and dividing unemployment projections and closings by the number of plants put "at risk" by the BPT regulations. All of these transformations have been discussed, but the use of sales of direct dischargers to normalize these variables requires further explanation.

We sought a normalizing measure that was available in our data sources and that would cause the least distortion to the relative value of the independent variable of interest. Annual sales possessed both of these characteristics. Using another available variable, such as the number of plants, would have biased the independent variable upwards for subcategories with a few large plants relative to those with many small plants. In addition, theory identified the number and size of plants being regulated as a possible explanatory variable; using that variable to normalize other variables would have confounded their separate effects. Sales, on the other hand, was not singled out as a variable of particular interest in explaining bureaucratic outcomes.

Nevertheless, annual sales for the plants in the subcategory would not be an appropriate normalizing measure because a proportion of sales can be attributed to plants sending their wastes to municipal treatment facilities; these plants are exempt from the BPT regulations. Because this proportion varies across subcategories, we scaled down subcategory sales by the percentage of plants discharging directly to waterways.

We encountered some special problems in converting data from one measure to another because of the obscurity of some of the measures. Data for dairies were sometimes in milk equivalents and sometimes in gallons; leather tanning data were in hides and weight; and timber data, in cubic feet and pounds. Conversion was made even more difficult because many of the data in the economic analysis reports were drawn from the 1967 *Census of Manufactures*, while the base year for our data base was 1972.

*The Aggregation Problem*

Aggregation problems arise in using aggregate data to describe or analyze features of individual entities that are only part of the aggregate. External signals theories describe the behavior of individual firms, yet the data reflect aggregations across these firms. If all firms could be considered identical, then the problem would disappear; but aggregate behavior may well differ from a summation of individual behavior. In this case, we are faced with problems in interpreting the results of analysis with data applicable to a group, as well as with the more mundane problem of aggregating data to the appropriate levels.

The huge number of subcategories created by the EGD for purposes of writing specific effluent standards, coupled with the EAD's relatively broad approach to describing the industry and projecting economic effects, created the need for much aggregation. For instance, regulations on the fruits and vegetable industry (phase II) covered up to fifty-eight commodity-specific subcategories, such as tortilla chips and ethnic vegetables, with further differentiation by size of plant. In contrast, the economic analysis report deals with the industry in terms of six SIC codes that cover fruit and vegetable canning, freezing, and dehydrating, while making an effort to exclude apple, citrus, and potato subcategories regulated under fruits and vegetables (phase I). In this case, we had little choice but to aggregate all of the subcategories to the industry level.

To aggregate the data for statistical analysis, we developed rules of thumb that related logically to the specific variable being aggregated. For many variables, such as projected plant closings or unemployment, we had only to add them for each of the finer subcategories contained in the aggregate category used for analysis. The level of detail the EAD had chosen to present the economic impacts of the regulations usually determined the subcategory aggregations used in our data base. Thus, the "adding up" rule of thumb was rarely used with economic variables, although raw waste and flow data, which were often available for every EPA subcategory, were simply added together for the subcategories included in one of our aggregate categories.

Aggregating the dependent variables in our data base required more complicated rules of thumb. Percentage change values were weight-averaged based on subcategory production shares because the standards were written in terms of effluent per unit of production. The percentage removal and concentration variables were aggregated using raw waste loads and flows, respectively, as weights. These weights were chosen because raw waste loads were used to calculate percentage removal values for each subcategory and flows were used to calculate concentration values for each subcategory.[22]

For the results of our study, the consequences of aggregation are unclear. Decisions by individual firms to oppose weaker standards and the resources made available for this purpose could only be reasonably explained by data on single firm characteristics. Using industry or even subcategory profit rates as a surrogate might prove to be a poor predictor of firm behavior. We are not trying to explain the decisions of individual firms, however. Rather, we focus on the consequences of those decisions

[22] These transformations were aided by work done by Patterson (1976).

and the decisions made by EPA officials and by other interested parties. In fact, to the extent that EPA personnel view the firms in a subcategory as homogeneous and gauge their decisions according to subcategory rather than firm characteristics, aggregate measures may be more appropriate predictors of regulatory outcomes. Moreover, a single firm's decision to oppose a regulation may be dictated by public-good considerations in the sense that a decision to fight a particular set of standards affects all firms with plants in that subcategory. In such cases, trade associations may, and in fact did, take the brunt of the responsibility for responding to particular regulations. Thus, the aggregate descriptors of industry characteristics may capture more fully the way a plant in a subcategory made its influence felt.

The proper degree of aggregation probably varies on a case-by-case basis. Nevertheless, both for industries with finely detailed subcategories and those with only general ones, decision making probably focused on major subcategories. Based on the comments made by industry, the rationale for standards made by EGD, the analysis of EAD, and a tracking of the changes in the standards over time, decisions apparently tended to be made for blocks of minor subcategories all at the same time.

Unfortunately, accepting our degree of aggregation to the major subcategory level as appropriate does not end the aggregation problem. Data limitations occasionally forced us to make use of industry-level data to describe plants in a major subcategory. This practice mostly affected the profit variable and occasionally the four-firm concentration ratio variable. The effects of this data mismatch may not be serious, however. First, many firms in an industry are integrated broadly enough to encompass most, if not all, of the products and processes covered by more than 1 of our 106 "major" subcategories. Thus, those firms' decisions may be based on a more aggregate profit measure than a profit rate applying to a single, major subcategory. Second, if we could not determine a profit rate or concentration ratio applicable to a subcategory, EPA personnel probably could not either. To the extent EPA made decisions based only on the industry-level information available to us, the industry figures may be reasonable ones.

## APPENDIX 5-C
*Modeling Categorization*

Chapter 4 stressed that, in addition to the stringency of a subcategory's standards, the decisions about which firms and plants to assign to different categories were important outcomes of the regulatory process. However, the discussion in that chapter also made it clear that we will not be testing hypotheses about categorization even though the simple model presented in chapter 4 did lead to a set of potentially testable hypotheses.

This decision was made because we found no meaningful way to measure the degree of industry categorization. Our simple theoretical model used the *number* of categories as a proxy for the degree of industry categorization. Examining the aggregate number of what EPA called subcategories within an industry category does not present any particular problems. Once a tracking system is devised to follow subcategories as they are split, combined, created, and dropped over the course of rulemaking, it is straightforward to count the number of subcategories at various stages to use as evidence about how categorization decisions were made.

However, for a variety of reasons, counting the number of subcategories created for each industry category would lead to potentially misleading or even meaningless hypothesis tests. In general, this result would occur because an explicit relationship between the number of subcategories and the stringency of subcategory standards (measured in terms of total abatement cost to firms in the industry) cannot be constructed; and the model of industry categorization rests heavily upon the existence of this relationship.

In the first place, it is risky to characterize the initial subcategorization decision made by the contractor as one based on costs. The contractor was supposed to account for differences in production processes or products, size, age, and other industry characteristics that might, from a technical (engineering) point of view, require different regulations. Sometimes these reasons were explicitly economic and sometimes not. Product differences were given as the rationale for creating most subcategories (such as those for taco chips and potato chips). However, the waste-water characteristics of many of these products could not have been much different. Therefore, abatement costs could not have been much different either.

There is a second, more important reason for not regarding many subcategorization decisions as being motivated by costs. EPA's decision to issue mass emission standards (in kilograms of pollutant per 1000 kilograms of production or material input), rather than standards expressed in percentage removal or concentration terms, is responsible for creating many subcategories. This decision meant that every industry segment with different types of products (for example, ducks, chickens, and turkeys in the poultry industry) required a separate set of standards.

Even plants producing similar products but using processes with different waste characteristics needed separate sets of standards. For example, six categories based on processes were established in the leather tanning industry, with promulgated standards for BOD ranging from 1.6 to 4.8 kg of BOD per 1000 kg of hide input. In every subcategory, however, the recommended technology was the same, as were the effluent concentrations specified in the Development Document: 75 mg/l for BOD, 95 mg/l for TSS, 211 mg/l for chromium, and 15 mg/l for oil and grease. Because the amount of waste generated per unit of hide differed greatly among categories, the discharge standards also differed. If the Effluent Guidelines Division had written the standards in terms of effluent concentrations, there would have been only one category in the leather tanning industry (ignoring possible subcategorizations based on size).

Although the above arguments convinced us that the relative number of subcategories at any stage was not a defensible measure of the degree of industry subcategorization, these arguments do not apply to *changes* in the number of subcategories over the course of the rulemaking process. However, this specification was also rejected because of its ambiguity. To understand this problem, consider two possible types of changes. First, new subcategories may have been created because some aspect of the industry had previously been ignored. In such a case, the number of subcategories increased and the total cost of regulation also increased (because more firms were covered by the regulation). This situation violates the assumption of the categorization model in chapter 4 about the inverse relationship between costs and the number of subcategories. Second, firms in a subcategory may have been excluded from a regulation, meaning that the total compliance cost declined. Again, the model's assumption is violated. Thus, it is apparent from these two examples that changes in the number of subcategories do not unambiguously measure changes in the degree of stringency and so a key assumption in the model is sometimes violated when the number of actual subcategories is used as a proxy for the degree of industry categorization. For this reason, we decided against empirically testing the subcategorization hypotheses.

134

# REFERENCES

Gianessi, L. 1978. "Estimates of National Water Pollutant Discharges by Polluting Sector: 1972" (Washington, D.C., Resources for the Future, unpublished).

———, 1976. "Appendix A2," in L. Gianessi and H. Peskin, eds., *The Cost To Industries of Meeting the 1977 Provisions of the Water Pollution Control Act Amendments of 1972,* report to EPA (New York, National Bureau of Economic Research, January).

Hausman, Jerry A. 1978. "Specification Tests in Econometrics," *Econometrica* vol. 46, no. 6 (November), pp. 1251–1272.

Patterson, James. 1976. "Directory of Federal and State Water Pollution Control Standards," document no. 77/06 (State of Illinois, Institute for Environmental Quality, October).

Van Note, Robert H., et al. 1975. "A Guide to the Selection of Cost Effective Wastewater Treatment Systems," Technical Report No. EPA-430/9-75-002 (Washington, D.C., U.S. Environmental Protection Agency, July).

# 6
# Analysis and Findings

In this chapter, the general hypotheses generated in chapter 4 are tested to determine which factors influenced the stringency of the Best Practicable Technology (BPT) standards set for different subcategories of firms. Environmental Protection Agency (EPA) decisions are examined to reveal the decision rules implied by the standards the agency issued.

The first question is whether some simple efficiency- and equity-based decision rules might explain the large variation in the BPT regulations, both across subcategories and from one stage of the rulemaking process to the next. Then, using multiple regression analysis and the data base described in chapter 5, supplemented where appropriate by other types of numerical analyses, the hypotheses that were generated in chapter 4 are tested. Finally, the picture of the effluent guidelines rulemaking process that emerges from the results taken as a whole is summarized.

## EFFICIENCY- AND EQUITY-BASED DECISION RULES

Tables 6-1, 6-2, and 6-3 describe several cost and other economic impacts projected to result from the BPT standards. Simply describing these consequences of the standards enables us to test directly whether EPA used some simple decision rules based upon efficiency or equity concerns.

TABLE 6-1.  MARGINAL TREATMENT COSTS OF BOD REMOVAL[a]
(dollars)

| Industry | Subcategory | Marginal cost[b] |
|---|---|---|
| Poultry | Duck—small plants | 3.15 |
| Meat packing | Simple slaughterhouse | 2.19 |
| | Low processing packinghouse | 1.65 |
| Cane sugar | Crystalline refining | 1.40 |
| Leather tanning | Hair prev. removed/chromium | 1.40 |
| Poultry | Duck—large plants | 1.04 |
| Leather | Save hair/vegetable | 1.02 |
| | Hair prev. removed | 1.02 |
| Meat packing | High processing packinghouse | .92 |
| | Complex slaughterhouse | .90 |
| Paper | Unbleached kraft | .86 |
| Leather tanning | Save hair/chromium | .75 |
| | Pulp hair/chromium | .63 |
| Poultry | Turkey | .60 |
| Cane sugar | Liquid refining | .51 |
| Paper | Paperboard | .50 |
| | Kraft—NSSC[c] | .42 |
| Leather | Pulp or save hair/no finish | .39 |
| Poultry | Further proc. only—large plants | .35 |
| | Chicken—small plants | .25 |
| Paper | NSSC—Ammonia process | .22 |
| Raw sugar processing | Louisiana | .21 |
| Poultry | Fowl—small plants | .20 |
| | Chicken—medium plants | .16 |
| Raw sugar processing | Puerto Rico | .16 |
| Paper | NSSC—sodium process | .12 |
| Poultry | Fowl—large plants | .10 |
| | Chicken—large plants | .10 |

*Source:* EPA development documents.

[a] Calculated by dividing incremental annual model plant costs to meet the proposed BPT standards by the incremental amount of BOD removed in the application of BPT technology.

[b] Measured in dollars per kilogram of BOD removed.

[c] NSSC = neutral sulfite semi-chemical.

## *Effluent Standards Based Solely Upon Efficiency*

If standards were set cost-effectively, marginal treatment costs would be equal across subcategories.[1] As table 6-1 indicates, however, estimates of marginal costs of BOD removal for various size model plants

[1] EPA's equalization rule for minimizing the total cost of compliance assumes that the damages per unit of pollution discharges do not vary across locations, clearly a false assumption. Because EPA did not know how damages varied across industry subcategories, however, the assumption is not unreasonable for purposes of analysis. The rule of equalizing the marginal cost of controlling each pollutant also simplifies the problem by ignoring the external benefits of controlling other pollutants with the same control efforts.

TABLE 6-2.  INCREMENTAL INVESTMENT COST PER KILOGRAM OF
BOD REMOVED

| Industry and subcategory[b] | Dollars of investment per kg of BOD removed per year[a] | |
| | Subcategory | Industry average[c] |
| --- | --- | --- |
| Builders' paper | | 1.34 |
| Poultry | | .97 |
| A | .70 | |
| B | .31 | |
| C | 59.09 | |
| D | 8.73 | |
| E | .44 | |
| Paper & pulp I | | .81 |
| A | 1.01 | |
| B | .89 | |
| C | .44 | |
| D | .52 | |
| E | .67 | |
| Leather | | .71 |
| A | .55 | |
| B | .87 | |
| C | .99 | |
| D | 1.02 | |
| E | .10 | |
| Sugar I | | .70 |
| A | .80 | |
| B | .30 | |
| Fruits & vegetables I | | .54 |
| A | 1.93 | |
| B | 1.31 | |
| C | .30 | |
| Fruits & vegetables II | | .24 |
| Meats I | | .19 |
| A | .35 | |
| B | .13 | |
| C | .23 | |
| D | .15 | |
| Sugar II | | .11 |
| A | .05 | |
| B | .30 | |
| Meats II | | .02 |
| A | 2.36 | |
| B | .03 | |
| C | .01 | |
| D | .01 | |
| E | (negligible) | |

[a] Based on the proposed BOD standards.

[b] Subcategories are identified by letters, rather than descriptive names. Roman numerals refer to EPA's designation of industry phase.

[c] Average for above industries = $.56 per kg BOD removed (standard deviation = $.42).

TABLE 6-3. PROJECTED ECONOMIC IMPACTS OF PROPOSED BPT REGULATIONS

| Industries[a] | Investment cost/sales of direct dischargers | Annual cost/sales of direct dischargers | Number of plant closures | Unemployment | Plant closures/number of plants at risk |
|---|---|---|---|---|---|
| | (———dollars———) | | | | |
| Glass—pressed & blown II | .029 | .011 | 4 | 850 | .25 |
| Leather tanning | .074 | .017 | 21 | 949 | .20 |
| Pulp and paper I | .23 | .094 | 6 | 480 | .07 |
| Builders paper | .24 | .164 | 4 | 350 | .19 |
| Cement | .024 | .007 | 0 | 0 | 0.00 |
| Sugar processing II | .034 | .015 | 11 | 1,750 | .26 |
| Poultry | .070 | .014 | 27 | 3,425 | .20 |
| Textiles | .054 | .016 | 142 | 2,702 | .44 |
| Sugar refining I | .008 | .003 | 3 | 300 | .16 |
| Meat packing I | .007 | .002 | 1 | 25 | .003 |
| Glass—flat I | .002 | .001 | 0 | 0 | 0.00 |
| Meat processing II | .001 | .002 | 16 | 800 | .04 |

Source: The Federal Register and EPA development documents, action memoranda, and economic analyses.
[a] Roman numerals refer to phase of industry.

in different subcategories range from $0.10 to $3.15 per kilogram of BOD removed. Thus, the standards could not have been cost-effective.

EPA may have focused more on the investment costs that groups of actual plants were expected to incur than on costs of model plants. Cost-effectiveness in this context implies equating the incremental investment costs per pound of pollutant removed at the subcategory or industry level. Table 6-2 lists some of these incremental costs. This measure also produces a wide range of costs. Both the marginal cost data in table 6-1 and the incremental investment cost data in table 6-2 indicate that considerations other than economic efficiency were driving the BPT rulemaking process.

*Regulation Based Only on Distributional Equity*

Although no generally accepted definition of equity exists beyond the vague generality of "treating equals equally," equity can be defined by borrowing the "equal sacrifice" principle from public finance. For example, EPA might have attempted to equalize costs by subcategory or industry, normalized by the number of plants or sales. A more practical approach would have been to manipulate subcategorizations and standards to avoid extremes—to make sure that no subcategory (or industry) suffered too greatly or too little. Given this view of equity, the regulators might have sought to avoid closing plants or causing unemployment, for example. The data in table 6-3, however, suggest that burdens were not equalized on normalized investment, annual costs, plant closings, or unemployment at the industry level. For instance, predicted closings vary from 0.003 percent for meat packing to 44 percent for textiles. Apparently a simple equity-based criterion, such as equalizing economic impacts across industries, cannot, in isolation, explain the variation in the BPT standards.

REGRESSION RESULTS

With no a priori expectations about the functional form of the model equations, a linear specification was chosen and estimated using ordinary least squares. A log-linear version of the model was also estimated, but as the results were not much different, they are not reported here. Also, the final set of explanatory variables that appears in the regression results (tables 6-4 and 6-5) was determined on the basis of theory, not goodness of statistical fit. Thus, the variables that the hypotheses identified as potentially important were retained in the econometric equations regardless of their associated *t*-statistics or contributions to *R*-squared. The statistical problems of heteroscedasticity and collinearity were addressed, however.[2] A diagnostic procedure available from the Statistical Analysis System (version 79.5), based on concepts of Belsley and others (1980), was used to assess the degree of collinearity. This program provides a numerical index that measures the conditioning of a matrix, that is, how close a matrix is to being singular. Following Belsley and others, a condition index of thirty or more was assumed to indicate the presence of collinearity. The procedure also provides a way of identifying which significant variables are contributing to the collinearity problem.[3]

Tables 6-4 and 6-5 present the results for the eight major regression equations, four for the BOD and four for the TSS pollutant standards, along with the hypothesized direction of influence for each variable in the regression. For each major pollutant, regressions (1) and (2) explain the level of the standards suggested by the contractors.[4] Regression equation number (3) explains the percentage change in the level of the standard from the contractor to proposed stages. Regression number (4) explains the percentage change from the proposed to promulgated stages. With the exception of the regression equation (2), explaining

---

[2] Heteroscedasticity causes problems because ordinary least squares estimates are efficient only if the residuals are uncorrelated. We followed the technique of visually examining the pattern of residuals for evidence of heteroscedasticity. Collinearity causes problems because the separate influence of collinear variables on the dependent variable may be masked. In addition, estimates of collinear variables may have large variances that make hypothesis tests unreliable. Such problems are traditionally identified by examining the zero-order correlation coefficient matrix for the explanatory variables. This technique can only deal with the relationships between two variables at a time. Three or more covariates may be collinear, however, even though any two are not. See David A. Belsley, Edwin Kuh, and R. E. Welsch (1980) for a discussion of problems with existing techniques and the analysis on which our collinearity diagnostics are based. See also *SAS Technical Report P-115 (1981)*.

[3] We found evidence of collinearity, but it mainly results from the relationship between *BODC* or *BODP* and the intercept. In addition, *CLOSE* and *UNEMP* are, not surprisingly, closely related ($r = 0.7$). Heteroscedasticity did not appear to be a problem for our data set.

[4] To simplify the presentation of the regression results, we omit those equations pertaining to the promulgated standards and those explaining changes in standards over the entire rulemaking process (that is, from the contractor stage to final promulgation). Little information is lost, however, because these regression equations are basically summaries of the process described by the other regressions. For example, the reasons for the percentage change in the standards from the contractor stage (C) to the promulgated stage (R) can be inferred by the two equations that explain the percentage change in the contractor to proposed (P) period and in the proposed to promulgated period. Furthermore, the explanation for the variation in the promulgated standards across subcategories can be derived from the results of the regression equation describing the contractor standards, plus the changes in the standards in the C to P period and in the P to R period.

variation in the TSS concentration standard, all regressions are significant (based on an *F*-statistic exceeding the critical value at the 5 percent level for a one-tailed test), they all have adjusted *R*-squared values exceeding 0.25, and they pass a Chow (*F*) test.[5]

The relatively poor results for the TSSC regression are not surprising. Because the TSS standards were designed to control a variety of inorganic waste streams, the applicable technologies vary more across industries than those for BOD treatment, where only organic wastes must be treated. We did not find a good measure to characterize that variation. In addition, the percentage removal BOD standards are bounded from above (at 100 percent), while the TSS concentration standards are not. Thus, the TSS standards should, and do, have more variability than the BOD standards (a coefficient of variation of 3.0 compared with 0.2). With this greater variability the *R*-squared should be lower for TSS.

## RESULTS OF HYPOTHESES TESTS

The econometric results in tables 6-4 and 6-5, supplemented by other types of statistical evidence, are the basis for testing chapter 4's hypotheses about regulatory agency behavior. The hypotheses are divided into two sets: those derived from the external signals model and those from internal, information-based theories of bureaucratic behavior. The strength of support for each hypothesis is evaluated by examining the statistical significance of the coefficient of the associated variable in the econometric equations and, when available, evidence supplementing the econometric results. For convenience, table 6-6 summarizes the hypotheses detailed in chapter 4.

### External Hypothesis Tests

In total, thirteen variables are used in the regression equations to test hypotheses derived from the external signals models of rulemaking. Each variable is associated with one of the parameters in the model.

*Number of Firms (Parameter m).* Hypothesis 1 states that industry categories with more firms, or plants, tend to receive more stringent

[5] A Chow test requires that the sample of observations be randomly divided into two subgroups. Separate regressions are run on each subgroup, and the hypothesis that the explanatory power summed over the two subsamples is higher than that of the entire sample is tested. If this hypothesis is rejected, then we conclude that the predictive value of the regression is not an artifact of the particular sample used and that the regression represents a stable relationship.

TABLE 6-4.   REGRESSION RESULTS FOR BOD STANDARDS

| Explanatory variables[a] | Dependent variables: | | | | Hypothesized sign | |
|---|---|---|---|---|---|---|
| | BSALESC (1) | BODC (2) | BCP (3) | BPR (4) | (1),(3),(4) | (2) |
| Constant | −33.06 | 7.73 | −57.17 | −49.10 | | |
| | (10.62) | (0.95) | (63.22) | (54.64) | | |
| CR4 | .01 | −.05* | .62* | .30 | + | − |
| | (.12) | (.01) | (.41) | (.53) | | |
| ROR | 1.78+ | −.29* | .66 | 5.28* | | |
| | (1.15) | (.10) | (3.42) | (2.65) | | |
| TAB | .003* | −.0003* | −.01 | −.01 | + | − |
| | (.002) | (.0002) | (.01) | (.005) | | |
| UNEMP | | | | −.05 | + | |
| | | | | (.66) | | |
| CLOSE | | | | −2.47 | + | |
| | | | | (26.33) | | |
| PI | | | | −3.20 | + | |
| | | | | (2.39) | | |
| INV | | | | 110.45* | + | |
| | | | | (55.58) | | |
| NC | | | .83 | −1.54 | + | |
| | | | (2.81) | (1.20) | | |
| | | | [NCBGP] | [NCBGCR] | | |
| NPAR | −.03 | .004* | −.01 | .02 | − | + |
| | (.03) | (.002) | (.08) | (.06) | | |
| PPAR | 10.53+ | −.07 | −23.77 | −5.45 | + | − |
| | (7.50) | (.67) | (17.74) | (15.72) | | |
| PSP | .04 | −.01+ | .16 | .43* | + | − |
| | (.07) | (.01) | (.18) | (.15) | | |
| RAWB | .06* | .002 | | | + | 0 |
| | (.02) | (.002) | | | | |
| FLOW | .02+ | −.002* | | | + | − |
| | (.01) | (.001) | | | | |
| DDQUAL | | | −53.05* | −20.94* | − | |
| | | | (15.05) | (12.13) | | |
| EAQUAL | | | 21.78 | −29.75* | − | |
| | | | (20.21) | (17.35) | | |
| TURN | | | 33.21* | 43.36* | + | |
| | | | (13.36) | (20.33) | | |
| ORDER | 2.56* | −.13* | −2.79 | .54 | − | + |
| | (.83) | (.07) | (2.95) | (2.17) | | |
| POLL | | | −5.26 | −13.28* | | |
| | | | (5.88) | (4.99) | | |
| BODC | | | 1.05* | 1.32* | + | |
| | | | (.35) | (.49) | | |
| BCP | | | | −.20* | − | |
| | | | | (.09) | | |
| $R^2$ | .61 | .50 | .49 | .73 | | |
| Adj $R^2$ | .53 | .41 | .32 | .55 | | |
| F | 8.02 | 5.24 | 2.87 | 4.25 | | |
| DF | 47 | 47 | 39 | 29 | | |
| CHOW $F^b$ | .79 | .64 | 1.28 | 2.39 | | |
| $CI^c$ | <30 | <30 | 47 | 40 | | |

TABLE 6-4.   (continued)

*Note:* Values in parentheses ( ) are standard errors. "*" and "+" denote significance at the 95 or 90 percent levels, respectively.

ᵃ See tables 5-2 and 5-4 for definitions of the explanatory variables.

ᵇ The *F* value used in the Chow test is calculated as:

$$F = \frac{[SSR_t - (SSR_1 + SSR_2)]/K}{(SSR_1 + SSR_2)/(N_1 + N_2 - 2K)}$$

where $SSR_t$, $SSR_1$, and $SSR_2$ refer to the sum of squared residuals for the entire sample and for each of the two subsamples respectively, $K$ is the number of regressors, and $N_1$ and $N_2$ denote the number of observations in each subsample.

ᶜ *CI* refers to the condition index.

standards. The variable *NPAR* (number of plants at risk, that is, incurring compliance costs) was used to test this hypothesis for the BPT rulemaking process. The econometric results only mildly support the hypothesis. Higher values of *NPAR* are significantly associated with tighter BOD standards at the contractor level (for the *BODC* measure, but not the *BSALESC* measure). Although in the other regressions the coefficients of *NPAR* are insignificant, they generally take a negative sign, indicating that standards are tighter and changes in standards (if positive) are smaller when *NPAR* is larger.

*Relative Signal Strength (Parameter e).*   Hypotheses 2, 3, and 4 relate relative signal strength of firms in the rulemaking process to the stringency of their standards. Hypothesis 2 states that active participation in the rulemaking process by firms (through the submission to the regulatory agency of well-documented comments in support of weaker standards) should increase their signal strength, resulting in weaker standards for them. The number of comments appearing in the *Federal Register* notices arguing for weaker standards (*NC*) was used to test this hypothesis. The coefficient of *NC* in the equation for changes in the TSS standards in the period from the contractor to proposed standards is both large and significant. None of the other three comment variable coefficients supports the hypothesis, however. These insignificant results held under a variety of specifications of the comment variable. For example, in explaining the change in standards from proposal to promulgation stages, the number of comments in the entire period from the contractor stage to final promulgation, as well as only the number of comments received in the second (P-R) period, were tried as independent variables.

The general insignificance of comments in informal rulemaking—also called notice-and-comment rulemaking—is troublesome and raises serious questions about the responsiveness of the process or about the specification of the comment variable. Counting the number of industry

TABLE 6-5.  REGRESSION RESULTS FOR TSS STANDARDS

| Explanatory variables[a] | Dependent variables: | | | | Hypothesized sign |
| | TSALESC (1) | TSSC (2) | TCP (3) | TPR (4) | (1),(2),(3),(4) |
|---|---|---|---|---|---|
| Constant | −1.41 (6.96) | −8.43 (42.05) | −19.94 (29.75) | 17.66 (29.19) | |
| CR4 | −.16* (.07) | −.48 (.39) | −.12 (.22) | −.62* (.24) | + |
| ROR | −.17 (.73) | 7.52* (4.43) | 5.62* (2.34) | 2.68 (2.59) | |
| TAB | .002* (.001) | .002 (.004) | .002 (.003) | −.003 (.003) | + |
| UNEMP | | | | 1.51* (.87) | + |
| CLOSE | | | | −24.74 (31.96) | + |
| PI | | | | −.34 (2.13) | + |
| INV | | | | 22.23 (18.86) | + |
| NC | | | 6.43* (2.29) [NCTGP] | −2.16 (.31) [NCTGCR] | + |
| NPAR | −.01 (.01) | −.05 (.07) | −.02 (.04) | .04 (.04) | − |
| PPAR | 14.03* (3.68) | 41.05* (22.22) | −27.29 (11.85) | 29.18* (14.02) | + |
| PSP | −.05 (.05) | −.31 (.29) | .01 (.14) | .05 (.16) | + |
| RAWT | −.001* (.0003) | −.001 (.002) | | | + |
| FLOW | .01* (.002) | −.01 (.01) | | | d |
| DDQUAL | | | −43.66* (8.31) | 12.38 (8.91) | − |
| EAQUAL | | | −14.65+ (10.34) | −15.46+ (11.08) | − |
| TURN | | | 68.25* (9.27) | 40.87* (13.29) | + |
| ORDER | 1.31* (.52) | 4.46+ (3.17) | −1.51 (2.08) | 2.97 (2.46) | − |
| POLL | | | .98 (2.13) | .45 (2.10) | |
| TSSC | | | .24* (.05) | .03 (.05) | + |
| TCP | | | | −.001 (.10) | − |
| $R^2$ | .53 | .11 | .64 | .43 | |
| Adj $R^2$ | .48 | .01 | .59 | .29 | |
| F | 10.47 | 1.11 | 11.23 | 3.00 | |
| DF | 85 | 85 | 81 | 71 | |
| CHOW $F^b$ | 1.31 | | .59 | 1.52 | |
| $CI^c$ | <30 | <30 | <30 | 32 | |

*Note:* See the footnotes to table 6-4 for an explanation of terms.

[d] Signs for *flow* hypotheses: >0 for column (1); <0 for column (2).

TABLE 6-6.  HYPOTHESES TESTED WITH THE REGRESSION MODELS

| Hypotheses |
| --- |

External factors

Hypothesis 1 — Firms in industry categories with more firms face more stringent standards.

Hypothesis 2 — Active commenting by firms to the regulatory agency in support of weaker standards results in weaker standards.

Hypothesis 3 — Higher predicted economic impacts (the level of inflation, employment losses, and plant closures) are associated with weaker standards.

Hypothesis 4 — The larger the percentage of an industry not in prior compliance with a regulation, the weaker the standards received by firms in the industry.

Hypothesis 5 — Firms in more concentrated industries receive less stringent standards.

Hypothesis 6 — Firms belonging to large or active trade associations receive weaker standards.

Hypothesis 7 — Higher annual compliance cost per plant or higher investment cost per plant at some fixed level of compliance result in weaker standards.

Hypothesis 8 — Industry categories dominated by small or old plants receive weaker standards.

Hypothesis 9 — Firms creating greater damage (if emissions are uncontrolled) receive less stringent standards.

Internal factors

Hypothesis 10 — Poorer technical support of the rules results in greater weakening of them.

Hypothesis 11 — The more turnover of important personnel during the process, the greater weakening of the rules.

Hypothesis 12 — Within a set of regulations issued under the same process, earlier rules will be less stringent than later ones.

Hypothesis 13 — If a proposed rule is made especially stringent during an early stage of the process, it will become less stringent at a later stage.

arguments that EPA chose to present in the *Federal Register* could have produced data filtered by EPA to justify its own position. Yet, EPA could omit mention of a significant comment only at some peril. The courts could, and did, overturn EPA decisions for failure to consider certain industry objections. Although the courts seldom cited industry's substantive arguments as grounds for remand, they often cited the "arbitrary and capricious" nature of a process that ignored industry objections.[6]

[6] In order to investigate directly whether EPA's reporting of comments was self-serving and selective, we read the *Federal Register* notices for the fruits and vegetables industry regulations and related these notices to the original comments. The notices showed that EPA treated the comments evenhandedly. The agency accepted many of industry's criticisms as being true and gave detailed explanations for why it disagreed with other comments. More important, the major points made in the original comments appear to be well represented in the *Federal Register*.

The specification of the *NC* variable is admittedly simplistic, although few realistic alternatives are available. *NC* ignores some important dimensions of the comment signal—its persuasiveness or timeliness, as well as the number of letters in which a given argument appeared. Still, short of performing a content analysis on every letter, no other proxy was available for this external signal.

Given these generally insignificant comment-variable coefficients and an understanding of the questionable quality of *NC*, what conclusion can be reached about the effect of industry comments on the stringency of the BPT standards? Either the comment variable did not adequately measure the influence of industry comments or else, in general, showering EPA with comments produced little change in the effluent standards. However, the specific objection raised in a comment (for example, that the proposed standard would cause the loss of forty jobs) may have been captured by another explanatory variable (for example, *UNEMP*). The result with respect to *NC* needs to be replicated in subsequent studies, but it at least suggests the possibility that regulatory agencies generally, and EPA in particular, rarely alter their standards in reaction to the volume of comments submitted in the informal rulemaking process.

Hypothesis 3 argues that, in an industry or subcategory for which the economic impacts of standards are high, the strength of opposition by firms will be enhanced, leading to weaker standards. The economic analysis reports forecast three primary economic impacts of the effluent guidelines: plant closings, employment losses, and product price increases. The variables *CLOSE, UNEMP,* and *PI* were used to measure these three impacts.

The regression results show no influence of either the plant closings or price increase projections on standard stringency. Many of the EPA project officers interviewed thought that, of these three variables, projected plant closings had the largest effect on the stringency of the BPT standards. Our results provide no support for this conventional wisdom. Closings were insignificant in regressions without unemployment as an independent variable. Projected unemployment appeared to have a strong and significant effect in weakening the TSS standards, but did not support the hypothesis for BOD standards. Nevertheless, unemployment projections appeared to be the most influential economic impact variable. Note, however, that unemployment and closure projections are closely related statistically ($r = 0.7$). Thus, it is difficult to distinguish the separate effect of these variables.[7]

---

[7] Two qualifications warrant mention. First, collinearity exists between *UNEMP* and *CLOSE* ($r = 0.7$). Second, because our economic impact data pertain to the proposed

Hypothesis 4 suggests that more homogeneous industry categories have greater signal strength and therefore receive weaker standards. Only one variable, *PPAR* (the percentage of plants in an industry at risk of bearing nonzero compliance costs), was available to measure the degree of compliance homogeneity. The results support the hypothesis that *PPAR* is inversely related to the stringency of contractor standards. All four coefficients of *PPAR* take signs consistent with this hypothesis, and three of the four are statistically significant. In contrast, the four coefficients of *PPAR* in the change equations show no clear pattern relating higher values of *PPAR* to the weakening of standards over the two periods of the rulemaking process.

*Probability of Firm Opposition (Parameter a).* Hypothesis 5, concerning the probability of firm opposition, argued that more concentrated industries receive less stringent standards. The data from the BPT process did not support this hypothesis. The coefficients of the industry concentration variable *CR4* took opposite signs for the BOD and TSS equations and were significant in one of the two contractor BOD equations. In addition, TSS standards were weakened less for subcategories with larger values of *CR4*; effects on BOD changes were insignificant. Because of these contradictory results, the industry concentration hypothesis was rejected, at least as relevant to the BPT rulemaking process.

According to Hypothesis 6, firms represented by more influential trade associations receive weaker standards and, by extension, a greater weakening of standards during the rulemaking process. The regression results show that trade association strength, as measured by *TAB* (trade association budgets), does appear to be a generally significant determinant of the stringency of contractor standards. All four coefficients of *TAB* in the BOD and TSS equations for contractors' standards show the expected signs and three of them are significant at a 95 percent confidence level. To take an example, the coefficient on *BSALESC* implies that an industry with a 10 percent higher trade association budget than another industry would have received a 5.5 percent weaker BOD standard at the contractor stage. In contrast, *TAB* shows no systematic influence on changes in the standards.

The credibility of these results as tests of the trade association influ-

---

standards, they could only explain changes in the proposed standards, not the earlier changes from the contractor to the proposed standards. Because the plant closing, unemployment, and price increase data were developed over the earlier period and passed on to the Effluent Guidelines Division, knowledge of these data could possibly have influenced the stringency of the proposed BPT standards. Our methodology is incapable of testing this particular possibility.

ence hypothesis depends, in part, on how closely our variable—the trade association's annual budget for 1972—correlates with "influence." On the one hand, EPA project officers, who were generally unfamiliar with the trade associations, might very well have treated representatives from the better known, better funded trade organizations with more deference. These organizations could have been expected to respond more quickly to information requests and various opportunities to influence EPA staff, as well as to employ more effective consultants. Trade associations become well funded because they are effective and represent many clients. With more clients, the pool of ideas and expertise available to structure trade association objectives is also larger. On the other hand, effective lobbying can sometimes be carried out by one knowledgeable, persistent person, especially when given staff support by firms in the industry. This conjecture, if true, would cast doubt on the usefulness of *TAB* as a measure of trade association influence.

These concerns about the accuracy of *TAB* as a proxy for trade association influence made the strong results linking higher values of *TAB* to weaker contractor standards even more surprising. These results suggest that the contractors and EPA, because it accepted the initial contractor standards as its own, treated industries with strong trade associations more leniently. Trade association activities themselves, however, may not necessarily have caused less stringent contractor standards to be issued. *TAB* could well be correlated with some unmeasured relevant characteristic of these industries. Nonetheless, as a predictor of which industries will receive weaker standards, trade association budgets appear to be important.

*Compliance Cost (Parameter r).*   Hypothesis 7 argues that discharging firms with higher annual compliance costs per plant or higher investment costs per plant generally receive weaker standards than those with lower annual or investment costs. Two variables were used to test this hypothesis for the BPT rulemaking process: *ANNUAL* (projected annualized costs to meet the proposed standards per dollar of sales of plants at risk) and *INV* (projected investment costs to meet the proposed standards per dollar of sales of plants at risk).[8]

Hypothesis 8 postulates that small plants tend to receive weaker standards. The *PSP* (percentage small plants) and *NSP* (number of small plants) variables for the BPT process were used to test this hypothesis.

Hypothesis 9 states that firms with higher emissions before control are likely to receive less stringent standards because they face higher

---

[8] Chapter 5 describes the rationale behind these particular choices for the normalization of annualized costs and investment costs.

compliance costs to meet any standard (that is, their compliance cost curve is relatively high). For the BPT rulemaking process, chapter 5 identified the raw waste load ($RAWB$ for BOD and $RAWT$ for TSS) and the waste-water flow ($FLOW$) as reasonable proxy measures for the relative heights of the compliance cost curves for firms in different subcategories. Thus, the model implies that both of these variables are negatively related to the stringency of the standards in that they are positively related to compliance cost, which, in turn, is inversely related to stringency of standards.[9] Tables 6-4 and 6-5 summarize the effects of $FLOW$, $RAWB$, and $RAWT$ on the levels of the BOD and TSS standards suggested by the contractors. Because the values of $FLOW$, $RAWB$, and $RAWT$ were determined by the contractor at the first stage of the rulemaking process, they could have affected the level of the contractor standards but not subsequent revisions in the standards.

What do the results reveal about all three of these hypotheses? Because $ANNUAL$ and $INV$ are highly collinear variables, only one was used in each regression equation. Tables 6-4 and 6-5 present the results for $INV$; the $ANNUAL$ results were found to be similar and therefore are not reported. Higher investment cost did have the expected significant association with weaker BOD standards in the period between the proposed and promulgated stages. The same result was found for TSS, except that the coefficient was significant only at a confidence level somewhat below 90 percent. Thus, the results provide modest support for hypothesis 7.

Small plants also may face higher treatment costs than larger plants meeting the same standards (because of economies of scale in treatment). For this reason small plants may receive weaker standards than larger plants. This hypothesis (8) was tested with the percentage of small plants variable ($PSP$), which appeared in all regressions. It is a significant variable for explaining both the level of BOD standards (specifically $BODC$) and their changes (specifically $BPR$), but is poor at explaining the variability of TSS standards. The regression results for the number of small plants $NSP$ (which were nearly identical to those for $PSP$) are not reported. (For further discussion of the $PSP$ coefficients, see appendix 6-A.)

Hypothesis 8, that small plants receive less stringent standards, cannot be rejected. In particular, the percentage (or number) of small plants is an important factor determining the degree of special treatment received. Turning this conclusion around, if small plants make up a small percentage of their industry, they may be discriminated against by the

---

[9] Remember that from their definitions, less stringent standards correspond to higher values of $BSALESC$, $TSALESC$, and $TSSC$ but to lower values of $BODC$.

rulemaking process; without evidence on treatment cost differentials among small plants in different industries, however, we cannot be certain about this interpretation. In addition, the evidence suggests that the rulemaking process favors large plants in industries with a moderately high percentage of small plants—enough to result in special treatment for the industry but not enough to justify creating a separate subcategory for the small plants.

According to hypothesis 9, because of the waste-water engineering characteristics of BOD and TSS removal, the waste-water flow ($FLOW$) and raw-waste load ($RAWB$ and $RAWT$) variables should be better proxies for the relative heights of the compliance cost curves than are $ANNUAL$ and $INV$. The four $FLOW$ equations summarized in tables 6-4 and 6-5 are all supported by coefficients with the correct signs, and all but the $TSSC$ coefficients are significant at above the 90 percent confidence level. The implied flow elasticity of $BSALESC$ is 0.34, with an elasticity of 0.20 for $TSALESC$, suggesting that a 100 percent difference in the waste-water flow rates between plants in two different subcategories (which often occurs) leads to standards whose stringencies differ on the order of 34 percent for BOD and 20 percent for TSS. The magnitudes of these point estimates indicate that variations in $FLOW$ and, by inference, variations in the relative heights of compliance cost curves, were a major determinant of the stringency of the BPT standards.

For the BOD standards, the raw-waste load ($RAWB$) results in table 6-4 strongly support the hypothesis; however, the TSS results in table 6-5 contradict the hypothesis. Given the greater overall difficulty in explaining the widely different TSS standards, we are not surprised by the latter result. Because, on theoretical grounds, the raw-waste load may serve as a better proxy for the height of the compliance cost curve for removing BOD than for removing TSS, the results still provide some evidence in support of hypothesis 9.

*Profitability.* Finally, although not posed as hypothesis, the preregulation profitability of firms in an industry could have influenced the stringency of standards in either direction. On the one hand, EPA may have perceived more profitable firms as better able to afford the expense of opposing its regulations. On the other hand, more profitable firms may have been seen as better able to absorb the compliance costs.

The regression results for the coefficients of $ROR$ (rate of return on equity) strongly indicate that more profitable industries receive weaker standards. All four of the coefficients on $ROR$ in the contractor equations take signs consistent with a weakening effect of higher rates of return, and three of them are statistically significant. In addition, all four coefficients in the change equations suggest that subcategories with higher rates of return receive more relaxation of their standards over the rulemaking process (two of them are significant). These results provide fairly convincing evidence that EPA gave more profitable firms weaker standards. Our theory attributes this behavior to the agency's desire to avoid the high level of opposition expected from such firms. Of course, other theories might offer different explanations for the observed effect of profitability. However, our results do not support the "compliance cost absorption" hypothesis.[10]

## Tests of Internal Hypotheses

*Information Quality.* Hypothesis 10 suggests that when regulated firms dominate the participation in the rulemaking process, as in the BPT process, better technical support of early versions of the regulations leads to less weakening of them over the course of the rulemaking process. Two dummy variables were used to measure the quality of technical support of the BPT regulations: *DDQUAL* (the quality of the development documents) and *EAQUAL* (the quality of the economic analysis reports). Because they measure the technical quality of supporting documents prepared after the contractor standards were issued, *DDQUAL* and *EAQUAL* could only have affected the revisions in the standards over the two periods in the process. Thus, they appear only in the four equations describing changes in the standards.

For both information-quality variables, three of their four coefficients provided significant support for the hypothesis. Given the crudeness of these measures of information quality, the results provide surprisingly strong support for the importance of the quality of technical information as an influence on the stringency of regulatory standards. The magnitudes of the coefficients are also surprising. For example, interpreting the coefficients of *DDQUAL* as the best point estimates of the impact on the stringency of standards from changing a low-quality development

---

[10] The appropriateness of *ROR* as a test of the competing profitability hypotheses depends upon the validity of using the aggregate industry profit measure as a proxy for individual firm profit rates (a necessity in this study), even though the profitability hypotheses apply to firms only. At the industry level, the distribution of compliance costs across an industry's firms, each with different profit rates, is one critical determinant of actual industry opposition. To assume that this distribution is similar across all of the regulated industries is probably not reasonable. Nevertheless, EPA's perceptions may not have been sensitive to the pre- and post-regulation distribution of such profit rates but may instead have focused more on the aggregate measure.

document to a high-quality one, the results indicate that subcategories with high-quality documents will have their BOD rules weakened 33 percent less in the first rulemaking period (and their TSS rules weakened 44 percent less) than subcategories with low-quality documents. Given that the mean values of *BCP* (the change in the BOD standards from the contractor to the proposed stage) and *TCP* (the same for the TSS standards) are 8.3 percent and 16.4 percent, respectively, these point estimates imply that high-quality development documents have a sizable effect on the stringency of standards.

Hypothesis 11 states that turnover of important personnel during the rulemaking process causes standards to be weakened when industry pressure predominates. Because technical project officers were among the most critical EPA employees involved in the BPT process, we used the dummy variable *TURN* (indicating whether the project officer for a subcategory changed during the process) to measure the turnover of key personnel during both periods of the rulemaking process. In all four change equations, *TURN* was highly significant, providing strong support for the hypothesis. Again referring to the *BCP* and *TCP* equations, the coefficients of *TURN* suggest that project officer turnover caused the BOD standards to weaken 33 percent more and the TSS standards to weaken 68 percent more than for subcategories without turnover.

When combined with the *DDQUAL* and *EAQUAL* results concerning the effect of information quality—in those cases written information as opposed to information known by one key individual—on the stringency of standards, better information allowed EPA to promulgate much more stringent standards.

*Learning.*    Based on a simple theory of learning by both regulator and regulatee, hypothesis 12 predicts that rules issued early in the regulatory process will be less stringent than those issued later. The *OR-DER* variable (the chronological order of promulgation) was used to capture this effect for the BPT process. Based on the regression results, the hypothesis can be rejected. In fact, the order of promulgation seems to work in the opposite direction; BPT rules issued by the contractor were less stringent the later they were issued. This result is based on the significant results applicable to three of the four contractor level standards. *ORDER* also did not significantly affect changes in the regulations during either period of the BPT rulemaking process.

Rather than supporting the organizational learning hypothesis, the *ORDER* results are more consistent with Downs' "law of increasing agency conservatism." From anecdotal knowledge of the process, we can offer several other explanations for the effect of *ORDER* on the contractor standards. During the initial stages of the BPT process, EPA was under strong pressure from industry, through both administrative and legal channels, to relax the standards. Industry's initial cascade of objections and lawsuits could have moved EPA to ask its contractors to take more lenient positions in recommending standards. Alternatively, contractors responsible for the later Phase II industry studies may have become sensitized on their own by experiences with the Phase I studies. Or perhaps the contractors were already trying to improve relations with the firms who would be needing help in implementing the BPT regulations in the future.[11]

*Dynamic Effects.* Hypothesis 13 suggests that if a rule under consideration is made especially stringent during an early stage of the process, it is likely to become less stringent at a later stage. This hypothesis provides the rationale for the expectation that tighter contractor (or proposed) standards would be associated with greater weakening, or less tightening, of the standards over the period from contractor standards to proposal (or from proposal to promulgation) and that greater weakening of the standards in the contractor to proposal period would be associated with greater tightening or less weakening in the following proposal to promulgation period.

The explanatory variables used to test these hypotheses were *BODC, BODP, TSSC, TSSP, BCP,* and *TCP.* The coefficients of the three variables in the BOD equations were significant at the 95 percent level. For TSS, the contractor's standard appears to positively and significantly influence the change in standards from the contractor to proposal levels, contrary to expectation; the other TSS effects are insignificant. Appendix 6-B provides another test of the hypothesis that a weakening of the standard in the contractor to proposal period causes a tightening of standards in the proposal to promulgation period.[12]

The two pieces of evidence taken together provide some evidence, at least for the BOD standards, that a weakening in a standard in the first

---

[11] An additional possibility is that EPA took a tougher stand with the subcategories regulated first because they were perceived to be, by congressional and EPA design, the most damaging water polluters. This hypothesis, however, appears to be inconsistent with the fact that *ORDER* did not influence changes in the standards. If EPA were trying to be tough on its major polluters, this attitude probably would not have been confined to the setting of standards at the contractor stage.

[12] Actually, the econometric formulation allows the weakening during the C-P period to cause a lesser weakening in the P-R period, whereas table 6-B1 examines whether the weakening in the C-P period causes a tightening (not just a lesser weakening) in the P-R period.

contractor to proposal period tended to result in less weakening, or even a tightening, of the standard in the subsequent period from proposal to promulgation.


## FINDINGS

Having confronted the fourteen hypotheses set forth in chapter 4 with the data from the BPT rulemaking process, what picture do the results paint of the factors influencing the setting of BPT standards within EPA? Or stated another way, what implicit decision rules do they reveal?

Simple rules based on either the concept of economic efficiency or the goal of distributional equity did not dominate the rulemaking process. All of the several possible definitions of efficiency and equity that were tested provide poor explanations of the large variations in standards across industry subcategories. EPA's standards were probably influenced by many factors simultaneously. These factors may be grouped into five factors, each measured by several variables in the econometric model.

First, the results indicate that EPA did adjust the stringency of regulations across industry subcategories to account for differences in compliance costs among firms (more technically, differences in the relative heights of the abatement cost functions among subcategories). This finding supports the commonly held view that the agency took into account compliance costs when setting its effluent guidelines. All four measures of relative compliance costs—INV, PSP, RAW, and FLOW—generally show significant coefficients with signs consistent with the compliance cost hypotheses for each variable.

Table 6-7 provides an illustration of the magnitudes of the influence of each explanatory variable on the final BOD standards promulgated (BSALESR). Because BSALESC's adjusted R-squared was higher than for BODC, the table uses BSALESC as the measure of the stringency of the contractor standards. The cumulative effects (caused by differences in the values of each of the explanatory variables among firms in different subcategories) are calculated from the coefficients in the BSA-LESC, BCP, and BPR columns of table 6-4. These cumulative effects reflect the combined influence of the variable at the contractor stage, plus the changes in the proposed and then the promulgated standards.

Column two provides the elasticities of the final BOD standards (BSALESR) with respect to each explanatory variable. For PSP, RAWB, and FLOW, the elasticities are all approximately one-third; the elasticity for INV is 0.11. In order to provide a better sense of the combined

TABLE 6-7. CUMULATIVE EFFECTS ON FINAL BOD STANDARDS OF
EACH EXPLANATORY VARIABLE
(percentage change)

| Variable | Cumulative effect on *BSALESR* per one unit difference | Cumulative effect on *BSALESR* per one percent difference | Cumulative effect on *BSALESR* per one standard deviation difference |
|---|---|---|---|
| CR4 | 0.99 | 0.40 | 22.31 |
| ROR | 18.70 | 1.25 | 33.85 |
| TAB | 0.00 | 0.08 | 0.53 |
| UNEMP | I | I | I |
| CLOSE | I | I | I |
| PI | I | I | I |
| INV | 110.45 | 0.11 | 23.19 |
| NC | I | I | I |
| NPAR | I | I | I |
| PPAR | 21.21 | 0.08 | 6.99 |
| PSP | 0.86 | 0.41 | 22.85 |
| RAWB | 0.41 | 0.36 | 54.45 |
| FLOW | 0.14 | 0.34 | 78.87 |
| DDQUAL | −73.99 | NM | NM |
| EAQUAL | −7.97 | NM | NM |
| TURN | 76.57 | NM | NM |
| ORDER | 14.68 | NM | NM |

*Note:* I means all coefficients are insignificant at 90% confidence level; NM means not a meaningful calculation.

impact of the variability in each explanatory variable and the responsiveness of final BOD standards per 1 percent variation in each variable (as measured by the elasticities), column three gives the percentage differences in final standards caused by a one standard deviation difference in the explanatory variables across subcategories. For all four measures of relative compliance costs, a one standard deviation difference leads to at least a 22 percent difference in the final BOD standards (and for *FLOW*, more than a 78 percent difference). These statistics demonstrate the sizable influence of the compliance cost factor in the EPA standard-setting process.

Second, information quality is the other major factor shown to create differences in the stringency of standards across polluting firms. In particular, the quality of the development document analysis leads to BOD standards 74 percent more stringent for high-quality than for low-quality documents. (For the TSS standards, the best point estimate is 31 percent.) To provide a further sense of the strength of this information-quality effect, column one of table 6-7 implies that a subcategory's compliance investment to sales ratio (*INV*) would have had to be 730 percent higher to completely offset the effect of a high-quality development document.

The results indicate that the quality of the analysis in the economic analysis reports was also influential in setting effluent guidelines, but much less so (an 8 percent increase in stringency due to *EAQUAL* as compared to 74 percent for *DDQUAL*). This result is consistent with the primarily null-effect results for the economic impact variables (*CLOSE, PI, UNEMP*) that were published in the economic analysis reports. If the key statistics generated by the report did not appear to influence the standards much, then there is little reason to suspect that the quality of the supporting analysis would matter. This result is also consistent with the institutional organization of BPT standard setting at EPA. Although the Economic Analysis Division had some review and analysis responsibilities, primary responsibility for the standards was lodged in the separate Effluent Guidelines Division. The information-quality effect is further supported by the project officer turnover results to the extent that turnover measures a loss of information held by the project officer. For subcategories experiencing turnover of the project officer, BOD standards were 77 percent weaker than subcategories without project officer turnover. (The best point estimate for the TSS standards is 109 percent.) The BOD result is about the same magnitude as the effect of a high-quality development document (*DDQUAL*).[13]

Industry characteristics are the third factor identified as having a possible influence on the standards, in this case by affecting the likelihood, as perceived by EPA, that industry would resist costly regulations. The results for the three specific industry characteristic variables (*TAB, CP4,* and *ROR*) are mixed.

Trade association budgets, either through the direct influence of the activities they support or as a proxy measure for industry influence, show a sizable and statistically significant influence on both the BOD and TSS standards at the contractor stage. The *BSALESC* elasticity with respect to *TAB,* for example, is 0.55. Interestingly, this influence does not extend beyond the initial contractor stages to revisions in the standards over the rulemaking process. Industry concentration (*CR4*) showed the hypothesized positive correlation with weaker BOD standards, but for the TSS standards the relationship was reversed leading us to reject the industry concentration hypothesis. In contrast, both the BOD and the TSS results solidly support the conclusion that more prof-

---

[13] Several case studies of the effluent guidelines process in individual industries have commented on the importance of information quality. The Urban Institute (1977) report, in particular, emphasizes the hypothesis that high-quality analysis affects the stringency of agency rules.

itable firms received weaker standards. The elasticity of the final BOD standard with respect to $ROR$ of 1.25 is the highest cumulative BOD elasticity in table 6-7 (see column two). From column three, a one standard deviation difference in $ROR$ between subcategories leads to a full 34 percent difference in the stringency of their final BOD standards. For reasons that statistical correlations alone cannot pinpoint, such as EPA's expectation that more profitable industries will resist costly regulation more strenuously, profitable industries appear to have received much more favorable treatment in the EPA decision-making process.

The volume of formal industry comments appears to have a highly limited influence in the rulemaking process. Although the TSS coefficient on $NC$ shows industry comments to have a significant influence in weakening standards during the first comment period from the contractor to proposal stage, the BOD coefficient is of the same positive sign, but insignificant. Both coefficients of $NC$ for the second comment period, from proposal to promulgation, are negative, strongly suggesting the lack of influence of the number of industry comments in this latter period of the rulemaking process.

The analyses in two single-industry case studies of the effluent guidelines process draw similar conclusions and provide added insight into them. Gaines (1977) examined the effect of comments on the corn wet milling rulemaking process and concluded that the extensive commenting by that industry had little effect on their standards. He explains this lack of influence by noting that the industry felt it lacked the technical data to argue against the regulations on their merits. Of course, the industry could have decided for strategic reasons to withhold technical information. In any case, this lack of data explains why the corn wet milling industry provided "a combination of very specific but inconsequential comments and sweeping unsupported statements" (Gaines, 1977, pp. 837–838).

The Environmental Law Institute's (thereafter ELI) 1979 case study of the phosphate manufacturing effluent guidelines provides an interesting explanation for our empirical finding that comments in the second period carried little weight and for hypothesis 13 (one also offered by Gaines and weakly supported by our statistical results) that comments in the first comment period, if anything, were more influential than those in the second. ELI observed that the industry comments in the contractor to proposal period consisted of technical discussions and proposals that directly addressed the problem of drafting reasonable regulations, whereas in the period from proposal to final promulgation the commenters appeared to be focusing on legal questions in preparation for eventual litigation.

The economic impact projections the agency carried out are the final factor hypothesized to be a major influence in the EPA decision rule for setting effluent guidelines. The statistical results imply that in the period from proposal to final promulgation of the regulation, neither plant closing nor price increase projections affected the stringency of the rules. The TSS equation shows a significant positive relationship between unemployment projections and the weakening of standards, but the BOD equation indicates no relationship, so at most the claim that higher unemployment projections caused the agency to relax its standards has only weak support.

These results stand in marked contrast to the assertions made by several nonstatistical case studies of effluent guidelines rulemaking. Both ELI and the Urban Institute (1977) claim that plant closings were an important determinant of the stringency of the BPT regulations. On the basis of interviews, Burt (1977) also asserts that predicted plant closings were a key determinant of stringency, but his only evidence appears to contradict the assertion. EPA predicted more plant closings for the dairy industry than for beet sugar manufacturers, but the former were given tighter standards. Many of the participants in the rulemaking process whom we interviewed also perceived plant closing projections to have strongly influenced the EPA rules, but the statistical evidence reveals a more mixed set of decision rules.

These implied EPA decision rules for setting effluent guidelines illustrate the types of findings that the revealed preference approach to agency decision making can produce. The factors hypothesized by theory, and by participants in the regulatory process, to have influenced the rules were sorted into important and minimal influences. Some of the results are not surprising (for example, that relative compliance costs mattered greatly). The results provide estimates of the magnitudes of the importance of all the explanatory variables and imply that trade-offs made by EPA influenced the stringency of its regulations.

## APPENDIX 6-A
*Small-Plant Effects*

The results for the *PSP* coefficients require further interpretation because the data base contains observations for separate subcategories consisting only of small plants, as well as observations for subcategories combining both small and larger plants. Inspection of the data shows that small-plant subcategories are invariably subject to weaker standards; that is, the small-plant-only observations are responsible for at least part of the *PSP* effect. In this situation, the *PSP* value of 100 percent for a small-plant subcategory is associated with a weaker standard than that of a large-plant subcategory, which has a *PSP* value of zero. We do not know, however, if the *PSP* effect extends to subcategories with intermediate percentages of small plants.

Table 6A-1 reveals that the situation in which small and large plants are found in the same subcategory (that is, where *PSP* is between 0 and 100 percent) is more common; fewer than one-third of the industries in our sample included subcategories based upon size. Thus, the fact that the significant *PSP* results for BOD are obtained even with the great number of "mixed-sized" subcategories implies some support for the hypothesis that industries without small-plant subcategorizations (but with higher percentages of small plants) receive weaker standards than those with lower values of *PSP*. The conflicting TSS results may be explained by the mixing of the two types of situations in our data set.

Table 6A-1 examines the question of why small-plant subcategories were created for one-third of the sectors but not for the others. The theory in the appendix to chapter 4 generates the hypothesis that subcategories are created in industries in which small plants have higher compliance costs or are present in greater percentages than are small plants in other sectors. Table 6A-2 supports the latter part of the hypothesis because, with two exceptions, only sectors with a high percentage of small plants ($\geq$39 percent) received a size subcategorization. Not surprisingly, even with these two exceptions, the percentage of small plants in the group with size subcategorization is significantly different from the percentage of small plants in the group without size subcategorization.[14]

---

[14] Glass I is a statistical anomaly because of the two SIC sectors involved. One has only one small firm, while the other has many. For each SIC sector, the percentage of small establishments is

| | | |
|---|---|---|
| 3211 | Flat glass | .03 |
| 3231 | Products of purchased glass | .69 |

Also, only a small portion of SIC 3231 is covered under the glass I regulation.

TABLE 6A-1.   SIZE DISTINCTIONS IN GROUP ONE INDUSTRIES

| Industries with at least one size subcategorization | Industries without size subcategorizations |
|---|---|
| Dairy | Grain milling |
| Fruits and vegetables II | Cane sugar |
| Seafoods | Cement |
| Feedlots | Plastics |
| Leather tanning | Soap & detergent |
| Electroplating | Nonferrous metals |
| Utilities | Phosphates |
| Petroleum refining | Ferroalloys |
| Meat processing | Glass |
| Rubber II | Asbestos |
| | Timber |
| | Pulp & paper |
| | Builders' paper |
| | Iron & steel |
| | Organic chemicals |
| | Inorganic chemicals |
| | Fertilizers |
| | Textiles |
| | Fruits & vegetables I |
| | Rubber I |

*Source:* EPA development documents and the *Federal Register.*
*Note:* All Group One industries are represented in this table, not just the ones in our sample.

## APPENDIX 6-B
*Further Evidence of Dynamic Effects*

Table 6B-1, a cross-tabulation of changes in BOD standards over the two periods, suggests that the stages of the rulemaking process were, indeed, not independent and that the process had a moderating influence on the standards; that is, weakened standards were subsequently tightened and tighter standards were weakened much more often than could be expected if the probability of a change at the contractor to proposed stage were independent of the probability of a change at the proposed to promulgated stage. This conclusion is reached by noting that sixty subcategories that were made less stringent between the contractor and the proposed stages were made more stringent between the proposed and promulgated stages, while only thirty-four of these subcategories could have been expected to have their standards changed in this way. The sixty-nine subcategories with standards becoming first more, then less, stringent exceed the expected number of forty-seven.

TABLE 6A-2.  SIZE SUBCATEGORIZATION AND PERCENTAGES OF SMALL PLANTS BY SELECTED INDUSTRIES

| SIC number | Industry | Number of establishments with fewer than 20 employees | Total establishments | Percentage of establishments with fewer than 20 employees | Does size sub-category exist? |
|---|---|---|---|---|---|
| 2062 | Sugar I | 5 | 33 | 15.2 | N |
| 2062 | Sugar II | 15 | 77 | 19.5 | N |
| 2611–2631 | Pulp & paper I | 104 | 682 | 15.2 | N |
| 2016 & 2017 | Meats III | 190 | 652 | 29.1 | N |
| 3011 | Rubber I | 80 | 206 | 38.8 | N |
| 3211 & 3231 | Glass I | 630 | 943 | 66.8 | N |
| 3221 & 3229 | Glass II | 144 | 369 | 39.0 | N |
| 3292 | Asbestos I | 45 | 142 | 31.7 | N |
| 3241 | Cement | 27 | 199 | 13.6 | N |
| 2661 | Builders paper | 18 | 92 | 19.6 | N |
| 2819 | Phosphates I | 119 | 383 | 31.1 | N |
| 287 | Phosphates II | 749 | 1,233 | 60.7 | N |
| 2865, 2869, & 3312 | Organics I | 388 | 1,051 | 36.9 | N |
| | Mean | 193 | 466 | 32.1[a] | |
| 22 | Textiles | 784 | 2,000 | 39.0 | Y |
| 2091 & 2092 | Seafood I | 433 | 828 | 52.3 | Y |
| 3111 | Leather | 294 | 517 | 56.9 | Y |
| 2032–2038 | Fruits & vegetables II | 1,168 | 2,258 | 45.7 | Y |
| 2021–2024, 2026 | Dairies | 2,523 | 4,590 | 55.0 | Y |
| | Mean | 1,040 | 2,039 | 49.8[a] | |

*Source:* U.S. Bureau of the Census, *1972 Census of Manufactures,* vol. II, *Industry Statistics,* pp. 1-3.
[a]Difference in means of percentage of small plants is significant at the 5 percent level ($t = 2.25$).

TABLE 6B-1.   CROSS-TABULATION OF CHANGES IN THE STRINGENCY OF
STANDARDS
(number of subcategories)

| Contractor to proposed stages | Proposed to promulgated stages | | |
|---|---|---|---|
| | More stringent | No change | Less stringent |
| More stringent | 11 (23) | 9 (19) | 69 (47) |
| No change | 8 (21) | 43 (17) | 31 (43) |
| Less stringent | 60 (34) | 12 (28) | 59 (69) |

*Note:* Parentheses indicate the expected number, $E_{ij}$, of subcategories in each cell, assuming independence in changes between levels. If $N_i$ is the sum of the ith row, $N_j$ the sum of the jth column, and $N$ the total number of subcategories, then

$$E_{ij} = \frac{N_i N_j}{N}$$

*Source:* Resources for the Future, Federal Regulations Data Base.

# REFERENCES

Belsley, David A., Edwin Kuh, and Roy E. Welsch. 1980. *Regression Diagnosis: Identifying Influential Data and Sources of Collinearity* (New York, John Wiley & Sons).

Burt, Robert E. 1977. "Effluent Limitations Under the Federal Water Pollution Control Act," in *Decision Making in the Environmental Protection Agency, Case Studies,* vol. 119 (Washington, D.C., National Research Council, National Academy of Sciences).

Environmental Law Institute. 1979. *Three Case Studies in Environmental Regulation* (Washington, D.C., ELI, January).

Gaines, Sanford E. 1977. "Decision Making Procedures at the Environmental Protection Agency," *Iowa Law Review* vol. 62, no. 3 (February) pp. 839–908.

*SAS Technical Report P-115.* 1981. (Cary, N.C., SAS Institute, Inc.)

Urban Institute. 1977. "EPA's Development of Effluent Guidelines for the Beet and Sugar Processing Industry," in *Organization Analysis of the Regulatory Process: A Comparative Study of the Decision Making Process in the Federal Communications Commission and the Environmental Protection Agency,* prepared for NSF Grant No. APR 75-16718 (Washington, D.C., Urban Institute).

# 7
# Further Implications of the Research Findings

Having demonstrated how the revealed preference approach can be applied to the problem of understanding agency rulemaking behavior, the question remains: Is this approach worth the cost? Does the revealed preference approach, with its emphasis on costly and time-consuming data development, further understanding of rulemaking any more than traditional, alternative approaches, and does it provide more confidence in whatever understanding it brings? This chapter summarizes the strengths and weaknesses of this approach relative to others. Part of the basis for this evaluation will be the results of the statistical tests taken as a group instead of individually as in chapter 6. The chapter also demonstrates how the insights gained from this effort may apply to regulatory issues of current interest. Finally, we offer some suggestions for future research.

## EVALUATING THE REVEALED PREFERENCE APPROACH

Table 7-1 summarizes the proportions of coefficients in each of the regression equations in tables 6-4 and 6-5 that provide statistically significant support for the hypotheses identified in chapter 4. These proportions and the adjusted $R^2$ values in the table both show mixed results. Slightly more than half of the four BOD equations' coefficients support the hypotheses, and slightly fewer than half of the four TSS equations' coefficients provided statistical support. In no cases were coefficients for the same variables in the BOD and TSS equations both statistically

TABLE 7-1.  SIGNIFICANCE OF RESULTS

| Regression equation—dependent variable | Percentage of hypotheses not rejected | Regression significance (adj. $R^2$) |
|---|---|---|
| *Panel A: BOD equations:* | | |
| Contractor standards (lbs/$ sales) | 62.5 (5/8) | 0.53 |
| Contractor standards—percentage removal | 75.0 (6/8) | 0.41 |
| % change in standards—contractor to proposed | 33.3 (4/12) | 0.32 |
| % change in standards—proposed to promulgated | 47.1 (8/17) | 0.55 |
|     Average for BOD equations | 54.5 | |
| *Panel B: TSS equations:* | | |
| Contractor standards (lbs/$ sales) | 75.0 (6/8) | 0.47 |
| Contractor standards—concentration (mg/l) | 25.0 (2/8) | 0.01 |
| % change in standards—contractor to proposed | 41.7 (5/12) | 0.59 |
| % change in standards—proposed to promulgated | 29.4 (5/17) | 0.29 |
|     Average for TSS equations | 42.8 | |

*Note:* The numbers in parentheses give the ratio of the number of coefficients with significant signs (at the 90% level) in the regression equation divided by the total number of coefficients in the equation. Since the sign of *ROR* coefficient is theoretically indeterminate, the results for this variable are excluded.

significant but of opposite sign. The equations explaining the variations in levels of the contractor-recommended standards generally performed better than the equations explaining the changes in the standards over the two periods in the rulemaking process.

These aggregate results provide a joint test of both the theory and the empirical methodology. They offer reasonable, but not overwhelming, support for this model of the effluent guidelines rulemaking process. Better theory or refinements in the methodology might improve upon their explanatory power. Interpreted another way, any ambiguity in these results could be due to insufficient precision in the theoretical basis for the hypotheses rather than the power of the methodology.

This assessment of the performance of the methodology used in this study does not address its usefulness as a means for explaining how regulatory agencies make rules. Costs and the other available approaches are relevant considerations. First consider the costs. Chapter 5 describes the practical problems of finding data appropriate for explaining the character of the BPT regulations. Under the assumption that decisions at EPA were primarily guided by the agency's perceptions of relevant factors, such as profitability, contained in its support documents, even though in conflict with other measures of these factors used outside the agency, we believe that sufficient data of acceptable quality were available to allow meaningful use of this statistical approach. Even so, these data were still time-consuming (and thus expensive) to collect. The costs of collecting data for some other rulemaking processes would probably be prohibitive unless many of the data were available in agency publications.

## COMPARISON WITH CASE STUDIES

The main alternative to the statistical approach for understanding how regulatory agencies make rules is the case study approach. Certainly the two approaches are complementary. The revealed preference approach, however, offers several advantages over relying on case studies. First, because rulemaking processes are extremely complicated, close examination of one or a few examples can easily miss subtle, but important, influences that a statistical analysis might reveal. Sometimes the most obvious factors are less important explanatory variables than the hidden ones. Second, the immediately obvious factors may be idiosyncratic to a particular process, meaning that they lack significance as a general influence on many rulemaking processes. Without examining a large number of regulations, idiosyncratic factors are difficult to separate from the general influences on rulemaking.

Finally, even if the revealed preference and case study approaches yield the same conclusions about which factors explain the variability in the rules issued, the statistical approach provides a much greater level of confidence in those conclusions. For example, the results showing the importance of the quality of the information in background documents are more reliable because they were derived from a sample of more than 100 industry subcategories than would be those based on only 1 or 2 subcategories. In addition, other researchers can replicate statistical results much more easily than if the conclusions depend critically upon the judgment of an individual researcher.

Chapter 6 compared this study's statistical results with those from several case studies of the BPT rulemaking process. A closer look at those case studies will help identify the differences in the contributions of each approach. The available case studies of BPT rulemaking (Burt, 1977; Environmental Law Institute, 1979; Gaines, 1977; and Urban Institute, 1977) present a detailed description of specific rulemaking processes for one or two industries, identify the key issues surrounding these processes, and provide many enlightening anecdotes about them. But they do not analyze why firms in certain subcategories fared better than others. For instance, the Gaines study of rulemaking for the corn wet milling industry centers on normative matters—primarily the question of the most effective mode of decision making at EPA as judged by eventual court approval of EPA rules, as well as by several other procedural criteria. In contrast, Burt's study focuses more on the stringency question. He states that the dairy industry faced tighter standards than the beet sugar manufacturers. Unfortunately, the discussion ends after he inquires into the cost-effectiveness of the standards; factors that might have created this situation are left unexamined.

In spite of the normative and generally procedural concerns of these studies, they serve an important role in generating hypotheses about the factors that influence the rulemaking process. Although these hypotheses are generally ad hoc, they may still be testable. Furthermore, the case studies delve deeper into the details of the rulemaking process than is possible with more formal, and therefore more simple, models of the regulatory process and generate hypotheses that might otherwise have been overlooked. As one example, the Urban Institute study hypothesized that the 1971 effluent standards issued under the Refuse Act of 1899 (33 U.S.C. Section 407 et seq.) served as a guidepost for the contractors who recommended standards in the first stage of the BPT effluent guidelines rulemaking process underway in 1973.

The case study approach cannot easily provide solid or confident tests of its hypotheses, however. We examined the "guidepost" hypothesis by comparing standards set under the Refuse Act of 1899 with those set for identically named subcategories during the contractor and promulgation stages of the BPT process. Only seven of seventy-one comparable subcategories had identical standards, while the BPT process generally produced weaker regulations.[1] Thus, the earlier standards probably had little effect on the BPT standards. As another example of the difficulty of using case studies to test hypotheses, the Urban Institute report attempts to compare the decision-making process at the Federal Communications Commission and at EPA, but it concludes that any meaningful general propositions from the four case studies in the project were difficult to support with confidence.

Burt and Gaines, as well as Parlour (1981) in his study of effluent standard setting for the Canadian pulp and paper industry, all consider the role of "internal factors" in the rulemaking process. They cite three factors as problems: tight deadlines, poor interagency communication, and agency difficulties in acquiring and interpreting data. Gaines and Parlour imply that these factors could lead to weaker regulations.

Our data also could not address the first two problems because the

---

[1] We note one interesting fact—the number of subcategories for comparable industries more than doubled from the Refuse Act process to the BPT process.

same tight deadlines and organizational structure characterized standard setting for nearly all BPT industries (and all of those in our sample).[2] Only an analysis of a number of different rulemaking processes with different time constraints and organizational structures could yield information about the effect of those factors. To yield credible conclusions, case studies must consider a variety of different rulemaking practices.[3]

Our results line up closely with the third internal factor the case studies cite—the effect of specific measures of information quality on the stringency of regulations. The Urban Institute report in particular emphasizes the hypothesis that high-quality analysis affects the stringency of agency rules.

The revealed preference approach can be successfully and profitably applied to rulemaking, although many obstacles remain before agencies would want to modify their procedures based on the findings of revealed preference studies. Further research using these methods can extend and improve this methodology to provide clearer and more credible guidelines for future reforms of rulemaking.

## IMPLICATIONS FOR RULEMAKING REFORM

Although the positive question of understanding the implicit decision rules that characterize the technical rulemaking process interests many social scientists, the normative question of how best to design rulemaking processes to achieve the legislative goals of a regulatory program holds far more interest to policymakers. To design rulemaking process reforms, however, requires understanding how regulatory agencies set rules under existing procedures and predicting how those rules would change under alternative procedures.

This study improves understanding of the factors that influenced EPA's BPT standards and may serve as a useful basis for understanding the factors that influence most technical rulemaking by social regulatory

[2] EPA chose to construct the technical analyses for the first two industries (insulation fiberglass and beet sugar processing) in-house because most of the technical information was available from previous studies and the agency wanted to learn how to develop a useful technical analysis. Contractors produced the development documents for all subsequent industries regulated under the effluent guidelines process.

[3] Gaines (1977) examined one example of each of three decisionmaking modes at EPA: informal rulemaking, formal rulemaking, and quasiadjudicatory decision making. His methodological approach did not allow him to systematically analyze the factors affecting the EPA decisions. It is extremely difficult to draw any inferences from his case studies about the influence of time constraints and the pattern of interagency communication.

agencies. Perhaps more important, the study provides guidelines for applying the revealed preference approach to other examples of rule-making. To the extent that subsequent studies of different rulemaking procedures discover different decision rules for setting standards, they will be useful in predicting the likely impacts of alternative designs for the rulemaking process. The complexity of the rulemaking process is too great and the decisions reached through the process are too important to rely solely upon insights derived from case studies and informed judgment.

*Some Reform Proposals*

Although our findings about the structure of decision making within the BPT process can by no means be considered the final word on the subject, they at least suggest the desirability and some of the consequences of several regulatory process reforms, some of which are being debated today within Congress, the agencies, and the academic literature.

One possibility follows from the regression result that certain industries were able to influence the rulemaking process at the contractor stage, before the first comment period began. Rather than trying to eliminate this precomment period influence—a task unlikely to be successful if industry holds the necessary data—participation of other interested parties in this stage of the process could be made easier. Contractors could be encouraged to seek the involvement of environmental or consumer groups, as well as the opinions of less profitable firms and less active trade associations. The use of a second contractor study, paid for by EPA but undertaken by a public interest group, although a more expensive option, could answer the objection of those who fear that such groups have too little technical knowledge to play an effective role. The use of a separate contractor to develop a common data base for the two studies, from in-plant sampling or from other direct contacts with industry, might be a way to reduce the cost of funding two separate reports while limiting industry influence in a realistic way.

Another possible reform would be to make the standards more cost-effective than our results show them to be. A comparison of marginal compliance costs across industries would help to increase cost-effectiveness. Although EPA rules are being developed for a number of industries at the same time (as at other agencies), each rule is often thought of as being distinct from others rather than as part of a larger program. It might be preferable for those agency officials who are responsible for developing industry regulations to meet with one another to discuss the technological basis for the standards and the marginal costs such standards imply. Informal attempts could be made to narrow the spread between such costs across industries.[4]

A third reform is suggested by the finding that industry subcategory standards were weakened considerably when the project officer in charge of the rulemaking process for the industry was changed. Turnover causes a loss of institutional memory that, in turn, makes arguments for changing the standards more difficult to judge on their merits. The need for eliminating such an arbitrary influence on regulatory outcomes is obvious, but how to go about preserving institutional memory is more open to debate. Project officers could be asked to keep a standardized set of notes, or they could be assigned assistants who would also gain an understanding of the industry, fill in temporarily for departing project officers, and transfer knowledge to their replacements.

*Prerulemaking Negotiation*

In addition to providing suggestions for new reforms, our results are relevant for evaluating current ideas for regulatory reform. One of these proposals is for prerulemaking negotiations, a procedure involving close and early cooperation between an agency and the groups affected by its rulemaking. This procedure is expected to head off controversies and raise key issues early on, thereby moving the rulemaking process along faster and more smoothly. It might also help obtain agreements from all parties, thereby reducing the number of court challenges.

Although these outcomes could be realized in some simple rulemaking procedures, our results point out that the number of independent interests in a rulemaking procedure is indeed large. In the BPT case, the desire of firms to improve their competitive position vis-à-vis other firms in the industry, coupled with the extraordinary number of ways in which firms can perceive threats to their self-interests, created pressure for fine-tuning the regulations to suit each firm or homogeneous group of firms. In spite of the initial intention of some rulemakers at EPA to

---

[4] Some would view this approach as an improvement over setting benchmark marginal cost standards as required in setting Best Conventional Technology Standards under the FWPCAA of 1977 (that is, an industrial cost-effectiveness standard for BOD, set with reference to the marginal costs for treating municipal wastes). It avoids the arbitrariness of a single benchmark standard in favor of consensus standards that are arbitrary only to the extent that the standards depend on the particular mix of industries being simultaneously regulated. Moreover, this approach avoids the political and legal challenges that are an inevitable consequence of benchmarks with wide application and substitutes an informal, and perhaps less contentious, operating procedure.

issue several generic effluent standards that would apply to firms in different industries, a bewildering number of subcategories was created; occasionally, the plants in each subcategory were even given a tailor-made standard to fit their unique waste stream or economic character-istics. Firms used their small size, their large size, their product mixes, their location, and a host of other characteristics to seek special treat-ment from EPA.

The use of the regulatory process to protect or enhance a firm's own interests is an inherent feature of the rulemaking process. Industry di-visiveness on the lead phase-down rule revision is a more contemporary example of the pressures that can arise when one regulated group sees its government-created advantages in danger of being lost through fur-ther government actions. EPA regulates the lead content in gasoline according to standards on three subcategories: large refiners, small re-finers, and blenders. EPA was considering the elimination of these subcategories, an action that would have tightened the standards for small refiners and blenders. (See *Inside EPA*, June 11, 1982.)

The requirements of the federal water pollution control legislation and EPA's administrative decisions have probably acted to create a process in which opportunities for special treatment proliferated. How prerulemaking negotiation by itself would reduce these opportunities or channel more constructively the energies that took advantage of or created them is difficult to understand. How, for instance, can a pre-rulemaking negotiation satisfy all interested parties? Simply negotiating with trade association representatives or officials from the largest firms probably will not work. Trade associations tend to be good at pressing for broad process changes but not for changes that benefit one segment at the expense of another; the interests of large firms may not coincide with smaller ones. Furthermore, as the size of the group expands to accommodate diversity, the difficulties of fashioning rules agreeable to all parties become prodigious. What mechanism of conflict resolution (outside the market) can handle these difficulties?

Environmental groups also would presumably be present in these negotiations. The diversity of interests among these groups may be less than that of industry; however, with the number of such groups so large, only a comparative few could gain representation at the negotiating table. How should the interests of the other groups be represented?

Even if industry and environmental groups can be chosen to represent their diverse interests and can arrive at compromise, that the interests of the public will be met is by no means certain. The new source per-formance standards requirement on coal burning by electric utilities is the classic case of a negotiated outcome (although not EPA-directed bargaining) that runs counter to the public interest. Environmentalists gained scrubbers on new utility stacks in the East and Midwest, Eastern and Midwestern coal and union interests received a boost to the use of high-sulfur coal, and consumers were stuck with a bill of $320,000 per worker job saved per year (Portney, 1982).

Our results also suggest that the regulatory process is already being influenced early by factors that are formally outside of it. Our econometric tests highlight the influence of profitability and trade association budgets as identifying characteristics of industries receiving more lenient contractor standards. These factors were not criteria to be formally considered during this early stage of rulemaking; formal contacts with industry come later in the process. Therefore, although the influence of these factors is not direct evidence of early ex parte contact with industry, it at least suggests that EPA was sensitive to industry concerns at a very early stage. Because firms in more profitable industries and with larger trade associations fare relatively better, and such outcomes may be unacceptable on equity or other grounds, perhaps prerulemaking negotiation can help EPA obtain a more acceptable allocation of the costs of regulation.

## Senate Bill 1080

Senate Bill 1080 is the latest of many recent congressional attempts to overhaul the basic machinery regulatory agencies use to issue rules. Our results offer some insights into the likely consequences of three regulatory process reforms that S. 1080 contains. This bill was passed unanimously in March 1982 and, after lack of House approval, was reintroduced in January 1983.[5]

First, consider the consequences of enhancing the powers of judicial review, as the Bumpers Amendment would have required. The S. 1080 version of the Bumpers Amendment requires the courts to determine whether the factual support for a rule is "substantial" on the record provided, in contrast to the more permissive "arbitrary and capricious" standard that Congress apparently thought some federal courts were applying. To the extent that passage of S. 1080 would induce the courts to increase their level of scrutiny of contested regulations, the quality

---

[5] Besides the reforms discussed here, S. 1080 also contained the legislative veto proposal. Since the Supreme Court's *Chadha* decision quashed the legislative veto in June 1983, the propects for S. 1080 are uncertain. At a minimum, it will have to be repackaged before the Senate gives it full consideration. For further discussion of S. 1080 and regulatory process reform, see Magat and Schroeder (1984).

of information agencies generated to support their regulations would assume more importance. Our results show a strong direct relationship between the stringency of the BPT standards up to, but not including, the judicial review stage and the quality of background information EPA used to support its effluent standards. Enhanced power of judicial review would make information quality an even more important[6] determinant of the ultimate stringency of agency rules.

Second, the explosive growth in the number of separate statutory procedures for carrying out notice-and-comment rulemaking has convinced many members of Congress that some standardized features for all rulemaking are necessary. S. 1080 would require, as in the BPT process, an advanced notice of proposed rulemaking (ANPR) and a comment period before proposed rules are issued. Many other examples of informal rulemaking currently lack this early comment period. We found that revisions to an industry's BPT standards during the process took place both during the contractor-to-proposal period and during the proposal-to-promulgation period, although the number of comments had little to do with these changes. These findings suggest that an additional comment period may lead to the proposal of regulations that are significantly different from those that would have been proposed without an ANPR. The additional comments received are unlikely to cause these differences, however.

In addition, S. 1080 would require all comment periods to extend at least sixty days. Our results about the ineffectiveness of industry comments provided during the thirty-day comment periods of the BPT process, although far from definitive, at least raise the possibility that doubling the length of the comment period may have little additional effect on regulatory outcomes. In this case, the major effect of lengthening the comment period may be a delay in the rate of rulemaking rather than in the stringency of the standards agencies set. This consequence of delay would be all the more likely if additional standardized procedures contained in S. 1080 were added. Some of these procedures are the (limited) right to make oral presentations; the (limited) right to cross-examine witnesses; and the fifteen-day extension of the comment period, which would be automatically triggered by significant new information provided to the agency during the original sixty-day comment period. Of course, lengthening the comment period would also allow more time for external and internal factors to operate on the process. Our analysis did not address the importance of additional time.

Finally, consider S. 1080's regulatory impact analysis requirement for cost-effectiveness and cost-benefit analyses. This requirement is a modification of a central element of Executive Order 12291.[6] Our regression results conflicted with the observation made by several EPA project officers that plant closings projections were an important influence on the stringency of BPT standards. Although this apparent conflict remains unresolved, we did find some limited support for the influence of unemployment projections as a determinant of the stringency of standards.

To the extent that patterns of decision making in the 1980s will repeat those of the mid-1970s, EPA and other regulatory agencies would continue to factor economic impact projections, such as plant closings and unemployment, into their standard-setting procedures, especially for minor regulations that are not costly enough to trigger the regulatory impact analysis requirement. Even for major regulations that do require a regulatory impact analysis, gaining adherence to the new cost-effective procedures will require changing what may be a well-established pattern of behavior developed in the past decade. Additional resistance may arise because engineers with no professional training in the use of cost-effectiveness analysis write most of the technical regulations.[7] At least in the short run, the major effect of the regulatory analysis requirement may be to slow down the rate of rulemaking without producing more cost-effective regulations.

## FURTHER RESEARCH

A number of data development problems must be solved before all the hypotheses suggested by the available theories of rulemaking can be tested. For instance, obtaining data on the characteristics of groups of people who benefited from water quality improvement in different industries and relating differences in those characteristics to the variation in industry standards would be useful. We were unable to collect these data for the BPT process, but they may be more readily accessible for other rulemaking activities.

As another example, much further empirical examination is needed relating the political responsiveness of regulatory agencies to the political preferences of members of Congress whose constituents are most closely tied to the regulations in question. One extension of our study along

---

[6] Briefly, Executive Order 12291 empowers OMB to hold up any major regulation—that is, one with impacts exceeding $100 million a year—until it can be supported by a regulatory impact analysis showing that, subject to statutory restrictions, it is economically efficient.

[7] The marginal cost figures we provide in table 6-1 demonstrate how cost-ineffective the BPT rules are.

these lines would be to relate the congressional voting records of key subcommittee chairmen to the stringency of the BPT standards issued for industries located within their districts. The Weingast and Moran (1983) analysis of congressional control of the Federal Trade Commission provides an excellent example of statistical research on the response of regulatory agencies to the wishes of Congress.

Another profitable extension would be to investigate further whether the division of responsibility for standard setting is an important determinant of outcomes. Our results suggest that the group with primary responsibility for the BPT standards, the Effluent Guidelines Division, had a greater influence than the group responsible for the economic impacts, the Economic Analysis Division. The fact that these groups were in different divisions and under different assistant administrators may have played a role in determining the outcomes. Because economic impacts are now assessed within the Effluent Guidelines Division, it might be possible to determine if this merger of responsibilities has had substantive effects and, more generally, if the degree of centralization of standard-setting responsibilities is an important determinant of regulatory outcomes.

A final extension is an examination of the influence on regulatory outcomes of two competing theories of agency behavior—the external signals theory and the congressional dominance theory. Both theories can be used to interpret our results that relate to external signals, such as the effects of predicted closures, unemployment, and general price level changes. Congressional voting records, socioeconomic and water-quality characteristics of the congressional districts, and data on other variables that might underlie congressional preferences could be compiled. A comparison of the ability of these variables to explain regulatory outcomes with that of the variables we used might shed some light on this issue.

The existing theories of rulemaking also should be refined. In particular, the external and internal theories of rulemaking need to be integrated into a single theory that combines the best elements of both. This new theory should recognize that the bureaucracy and regulated groups have heterogeneous interests. Some work has explored these conflicting interests within a governmental bureaucracy, but little has been written on the conflicting interests within an industrial bureaucracy or among firms. For example, the differences in the interests of profitable firms with modern plants and of marginal firms with old plants may explain many regulatory outcomes. Our general model in chapter 4 provides a start in this direction but is too complex to be useful for generating testable hypotheses.

The scope of our research was purposely limited to studying regulatory actions from their inception within an agency to final promulgation. Future research should consider extending this scope to the postpromulgation phases. In particular, judicial decision making must be studied as an important part of an informal rulemaking process. The design and operation of the permitting and enforcement phases of the process also may affect the design of specific rules. Furthermore, the design and stringency of rules can affect how they are enforced in the field.

The depth, as well as the breadth, of our research, could usefully be extended. Our sample included twenty-three of forty-nine Group One industries. Increasing the number of sample industries would allow tests of industry-level hypotheses, including those explaining the variation in the degree of subcategorization within industries. Tests of subcategorization hypotheses, however, must await further progress in developing a theory that accurately captures the essence of the subcategorization decision problem. Tests of subcategorization hypotheses must also wait for the development of better empirical measures of the degree of industry subcategorization.

Finally, and perhaps most important, the statistical approach to understanding bureaucratic behavior should be extended to other agencies, as well as to other rulemaking procedures at EPA. These procedures need not be limited to technical rulemaking. Indeed, the theoretical models of chapter 3 may be applicable to other types of rulemaking procedures. More results that either support or contradict the findings of our BPT study would increase confidence in drawing general conclusions about the implied decision rules that characterize decision making within social regulatory agencies. The more procedures that are analyzed using the revealed preference approach, the richer will be our understanding of the rulemaking process. The mounting frustration with both the procedures and the decisions of regulatory agencies makes this agenda for future research a most important task.

# REFERENCES

Burt, Robert E. 1977. "Effluent Limitations Under the Federal Water Pollution Control Act," in *Decision Making in the Environmental Protection Agency, Case Studies* vol. 119 (Washington, D.C., National Research Council, National Academy of Sciences).

Environmental Law Institute. 1979. *Three Case Studies in Environmental Regulation* (Washington, D.C., January).

Gaines, Sanford E. 1977. "Decision Making Procedures at the Environmental Protection Agency," *Iowa Law Review* vol. 62, no. 3 (February) pp. 839–908.

Magat, Wesley, A., and Christopher Schroeder. 1984. "Administrative Process Reform in a Discretionary Age: The Role of Social Consequences," *Duke Law Journal* vol. 1984, no. 2, pp. 301–344.

Parlour, J. W. 1981. "The Politics of Water Pollution Control: A Case Study of the Canadian Fisheries Act Amendments and the Pulp and Paper Effluent Regulations, 1970," *Journal of Environmental Management* vol. 13, pp. 127–149.

Portney, Paul R. 1982. "How *Not* to Create a Job," *Regulation* vol. 6, no. 6 (Nov./Dec.).

Urban Institute. 1977. "EPA's Development of Effluent Guidelines for Beet and Sugar Processing Industry," in *Organization Analysis of the Regulatory Process: A Comparative Study of the Decision Making Process in the Federal Communications Commission and the Environmental Protection Agency*, prepared for NSF Grant No. APR 75-16718 (Washington, D.C.).

Weingast, Barry R., and Mark J. Moran. 1983. "Bureaucratic Discretion or Congressional Control: Regulatory Policymaking by the Federal Trade Commission," *Journal of Political Economy* vol. 91, no. 5, pp. 765–800.

# Index

178